DISCARDED

# Fate and Weathering of Petroleum Spills in the Marine Environment

A Literature Review and Synopsis

by
Randolph E. Jordan
James R. Payne

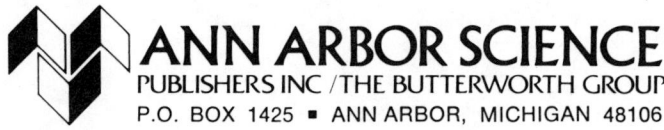

ANN ARBOR SCIENCE
PUBLISHERS INC / THE BUTTERWORTH GROUP
P.O. BOX 1425 ■ ANN ARBOR, MICHIGAN 48106

Copyright © 1980 by Ann Arbor Science Publishers, Inc.
230 Collingwood, P.O. Box 1425, Ann Arbor, Michigan 48106

Library of Congress Catalog Card No. 80-66473
ISBN 0-250-40381-1

Manufactured in the United States of America
All Rights Reserved

The work on which this publication is based was performed pursuant to Contract No. 03-7-002-35213 with the National Oceanic and Atmospheric Administration.

PREFACE

This manuscript was initiated as a literature review for Dr. John Calder of the National Oceanic and Atmospheric Administration as part of an extensive research effort to yield a thorough understanding of petroleum weathering in the marine environment. Computerized library searches were performed and condensed for extraction of pertinent references which formed a broad base of information for incorporation into the review. Other references were provided from the authors' own collection of reprints and from consultations with petroleum hydrocarbon scientists from industry and academia. Where omissions or oversights of other published works have occurred, the authors can only apologize and accept full responsibility.

As a guiding principle, the nature of this review was primarily determined by our desire to expose as much pertinent information as possible on the various chemical and physical alterations occurring to petroleum as a result of natural processes. Biological effects of petroleum hydrocarbon spills are not considered in detail.

The original draft of this text was reviewed by several prominent scientists from the field of petroleum hydrocarbon chemistry, and where appropriate, we have incorporated the comments or corrections supplied by Dr. Patrick Parker, University of Texas; Dr. Juanita Gearing, University of Rhode Island; Dr. Donald Mackay, University of Toronto and Dr. John Calder, NOAA.

The authors gratefully acknowledge support of NOAA under contract number 03-7-002-35213. We also wish to thank Mrs. Cheryl Fish and Miss Mikki LeMoine for the production of this review; Ms. Keta Hodgson, for text production/editing; and Miss Randee Luedecke and Ms. Kirstin Lanham for graphics support.

                                        Randolph E. Jordan
                                        James R. Payne
                                        May 5, 1980
                                        La Jolla, California

Randolph E. Jordan is an Associate Chemist, Division of Environmental Chemistry and Geochemistry, Trace Environmental Chemistry Laboratory, Science Application, Inc., La Jolla, California.

After receiving a BA in Biology from the University of California, San Diego, Mr. Jordan joined Science Applications, Inc., and has been working on environmental research dealing with petroleum and chlorinated hydrocarbon biogeochemistry in terrestrial and marine environments. He has also been responsible for methodology development programs on priority pollutant screening in wastewaters and other environmental media.

Mr. Jordan is presently the principal investigator of a collaborative project with Scripps Institute of Oceanography investigating the relative influence of microbial metabolism on the disposition of crude petroleum in a simulated marine environment. The project is part of a three-year program designed to provide a multivariant analysis of petroleum weathering which will assess the relative importance of various biotic and abiotic processes, and their associated environmental parameters, in the fate of petroleum introduced into the marine environment.

James R. Payne is Senior Chemist, Division of Environmental Chemistry and Geochemistry, Trace Environmental Chemistry Laboratory, Science Applications, Inc., La Jolla, California. He received his PhD in Chemistry from the University of Wisconsin--Madison, and a BA cum laude in Chemistry from California State University--Fullerton. He was a National Institutes of Health predoctoral fellow at Wisconsin and focused his research interests on apoenzyme/coenzyme interactions and the synthesis of coenzyme analogs, including $^{13}$C-enriched vitamin B-6, for $^{13}$C nuclear magnetic resonance studies of enzyme/coenzyme active site complexes. After graduate school, he received a Woods Hole Oceanographic Institution Post-Doctoral Fellowship, where he undertook research on the incorporation of petroleum hydrocarbons into marine shellfish, and on the persistence and metabolism of PCB in the water column of the North Atlantic.

Later research has centered on development of laboratory methods and large-volume seawater sampling systems for detection of trace-level organics, particularly petroleum hydrocarbons, in seawater and tissues of selected marine species. He was an invited scientist on the NOAA ship, Researcher, which investigated the IXTOC oilwell blowout in the Bay of Campeche, and is currently principal investigator of a three-year laboratory and field program designed to provide a multivariant analysis and computer model of petroleum weathering after release into simulated and natural marine environments. Dr. Payne has also conducted research on HPLC and glass capillary GC/MS techniques for the analysis of pollutants and their metabolites in animal tissues as well as the characterization of anthropogenic wastes in other environmental media.

CONTENTS

Introduction . . . . . . . . . . . . . . . . . . . . . . . . 1

I. Abiotic Factors/Processes . . . . . . . . . . . . . . 3

   Spreading . . . . . . . . . . . . . . . . . . . . . . 3
   Drift . . . . . . . . . . . . . . . . . . . . . . . . 6
   Spreading as Influenced by Detergents or
     Emulsifying Agents . . . . . . . . . . . . . . . . 9
   Evaporation . . . . . . . . . . . . . . . . . . . . 12
   Photooxidation . . . . . . . . . . . . . . . . . . . 23
   Dispersion . . . . . . . . . . . . . . . . . . . . . 32
   Dissolution . . . . . . . . . . . . . . . . . . . . 35
   Emulsification . . . . . . . . . . . . . . . . . . . 42
   Tarball Formation . . . . . . . . . . . . . . . . . 43
   Agglomeration, Sedimentation and Sinking of
     Petroleum After Release into the Marine
     Environment . . . . . . . . . . . . . . . . . . . 44

II. Microbial Degradation . . . . . . . . . . . . . . . 55

   Introduction . . . . . . . . . . . . . . . . . . . . 55
   Distribution and Types of Hydrocarbonoclastic
     Microorganisms . . . . . . . . . . . . . . . . . 55
   Petroleum Hydrocarbons Known to Be Oxidized and
     Rates of Utilization . . . . . . . . . . . . . . 59
   Factors Affecting Utilization and Rates . . . . . . 64
   Interference/Enhancement of Degradation Rates . . . 65
   Metabolic Pathways of Microbial Petroleum
     Degradation . . . . . . . . . . . . . . . . . . . 68

III. Physical and Environmental Interactions of Spilled
     Oil . . . . . . . . . . . . . . . . . . . . . . . 93

   Interactions with Shoreline Environments . . . . . . 93
   Oil in Near Shore Sediments . . . . . . . . . . . . 102
   Oil in Estuarine and Moderately Exposed Environments 104
   Oil Released in Arctic Environments; Oil and Ice/
     Snow Interactions . . . . . . . . . . . . . . . . 108

IV. Properties and Types of Crude Oils and Petroleum
    Products . . . . . . . . . . . . . . . . . . . . . 115

| V. | Approaches to Mass Balance Problems | 125 |
|---|---|---|
| | Open Ocean Mass Balance | 127 |
| | Mass Balance in an Estuarine Environment | 131 |
| | Mass Balance of Oil Released in Ice Covered Areas | 132 |
| VI. | Prospectus | 137 |
| References | | 141 |
| Index | | 167 |

## FIGURES

I-1. The four forces that act on an oil film . . . . . 4
I-2. Process vs time elapsed since the spill . . . . . 4
I-3. Wind pattern for March 17-April 10 from the French meteorological station 1 km north of l'Aber Wrac'h . . . . . . . . . . . . . . . . . 7
I-4. Predicted and observed location of the Orion buoys with forecast winds . . . . . . . . . . . . . . . 8
I-5. Observed and predicted location of the Orion buoys with observed (actual) winds measured at the site . . . . . . . . . . . . . . . . . . . . . . 8
I-6. Decrease in dissolved oxygen concentrations as a function of time . . . . . . . . . . . . . . . . 11
I-7. Decrease in the relative composition of hydrocarbons and heteroaromatic compounds . . . . . 13
I-8. Weathering rates of hydrocarbons from mousse . . . 13
I-9. Vapor pressure as a function of carbon number for alkanes, isoalkanes, cycloalkanes and aromatics 17
I-10. Percent of low-boiling hydrocarbons remaining in south Louisiana crude oil slick . . . . . . . . 21
I-11. Percent of aromatic hydrocarbons remaining in surface oil slick--first La Rosa spill . . . . 21
I-12. Optical density vs wavelength and oil film thickness . . . . . . . . . . . . . . . . . . . . 25
I-13. Hypothetical mechanism for sensitizer-induced free-radical oxidation . . . . . . . . . . . . . 27
I-14. Increase in oxidation products with time, as given by the IR absorption ratios of $C=O/CH_2$ . . . . 28
I-15. S and lens diameter $\delta$ vs irradiation time for light Tiajuana oil . . . . . . . . . . . . . . . 31
I-16. Crude oil residues--lens diameter vs irridiation time . . . . . . . . . . . . . . . . . . . . . . 31
I-17. Crude oil residues containing 1-naphthol; lens diameter vs irradiation time . . . . . . . . . 31
I-18. Natural log of solubility (as mole fraction) vs molar volume . . . . . . . . . . . . . . . . . 38
I-19. Natural log of normalized solubility (as mole fraction) vs molar volume . . . . . . . . . . . 38
I-20. Glass capillary FID gas chromatograms of seawater extracts and suspended particulate matter extracts collected simultaneously off Goleta Point . . . . . . . . . . . . . . . . . . . . . . 47
I-21. Glass capillary FID gas chromatograms of seawater extracts and suspended particulate matter extracts collected simultaneously off Corona del Mar . . . . . . . . . . . . . . . . . . . . . 48

| | | |
|---|---|---|
| I-22. | Glass capillary FID gas chromatogram and partial IR spectra of extracts from oil-kaolinite modeling experiments . . . . . . . . . . . . . . | 51 |
| III-1. | Surface oil cover at two tidal levels at Chedabucto Bay stations from 1970 to 1975 . . . | 94 |
| III-2. | Decrease in aliphatic hydrocarbon concentrations over time after the Amoco Cadiz oil spill--l'Aber Wrac'h sediment samples from Station A . | 99 |
| III-3. | Glass capillary FID gas chromatograms of the aliphatic fractions of hydrocarbons extracted from sediment samples immediately after the Amoco Cadiz and after seven months of weathering . . . . . . . . . . . . . . . . . . | 100 |
| III-4. | Decrease in aromatic hydrocarbon concentrations over time after the Amoco Cadiz oil spill--l'Aber Wrac'h sediment samples from station A . | 101 |
| III-5. | Glass capillary FID gas chromatograms of the aromatic fraction of hydrocarbons extracted from sediment samples immediately after the Amoco Cadiz spill and after seven months of weathering . . . . . . . . . . . . . . . . . . | 103 |
| III-6. | Flow of oil in rafted ice . . . . . . . . . . . | 113 |
| IV-1. | Boiling point range of fractions of crude and refined petroleum . . . . . . . . . . . . . . . | 116 |
| V-1. | Estimated fate of oil from Amoco Cadiz . . . . . | 126 |
| V-2. | Mass balance of untreated spill oil, April and December . . . . . . . . . . . . . . . . . . . | 128 |
| V-3. | Mass balance of treated spill oil, April and December . . . . . . . . . . . . . . . . . . . | 128 |
| V-4. | Trajectories and areas of the surface spillets and the dispersed oil (50-ppb contour) for the treated and untreated spill in April . . . | 130 |
| V-5. | Trajectories and areas of the surface spillets and the dispersed oil (50-ppm contour) for the treated and untreated spill in December . . | 130 |

## TABLES

| | | |
|---|---|---|
| I-1. | Percent loss caused by evaporation of No. 2 heating oil as a function of the field conditions | 15 |
| I-2. | Evaporative losses as a function of the Beaufort sea state | 18 |
| I-3. | Loss of volatile hydrocarbons from oils on water surface | 19 |
| I-4. | Density and kinematic viscosity of two weathered hydrocarbons in seawater and distilled water at $25^{\circ}$C | 22 |
| I-5. | Solubilities of aliphatic and aromatic petroleum hydrocarbons in seawater and distilled water at $25^{\circ}$C | 39 |
| I-6. | Additional solubility data on a variety of petroleum hydrocarbons in distilled water and filtered seawater | 41 |
| II-1. | Microorganisms capable of oxidizing/cooxidizing petroleum hydrocarbons and/or their derivatives | 57 |
| II-2. | Petroleum hydrocarbon compounds known to be oxidized (directly or co-oxidized) by hydrocarbonoclastic microorganisms | 60 |
| II-3. | Microbial co-oxidation of petroleum hydrocarbon compounds | 70 |
| II-4. | Examples of microbial n-alkane oxidations | 75 |
| III-1. | Chemical and physical characteristics of Bunker C oil in sediments at Chedabucto Bay | 95 |
| III-2. | An oil spill vulnerability index with particular reference to the Amoco Cadiz oil spill | 98 |
| III-3. | Concentrations of aromatic hydrocarbons in l'Aber Wrac'h, determined by GC/MS ERCO | 102 |
| IV-1. | The pour points for common crude oils and selected gas oil fractions | 115 |
| IV-2. | Analysis of typical Prudhoe Bay crude oil | 117 |
| IV-3. | Physical characteristics and chemical properties of several crude oils | 118 |
| IV-4. | Refinery fractions by hydrocarbon types from crude petroleum | 121 |
| IV-5. | Physical characteristics and chemical properties of two refined products | 122 |
| IV-6. | Viscosity alterations as a function of time and temperature | 124 |
| V-1. | Oil budget for Bouchard #65 oil spill, February 2-4, 1977 | 135 |
| V-2. | Weathered losses of oil in January 1977 | 136 |

# INTRODUCTION

The problem of oil pollution in various marine and estuarine environments has received considerable scientific attention with respect to the fates and effects of petroleum spills, as well as inherent toxicities to specific biological ecosystem components (i.e. planktonic communities) and individual species. It is most certain that the incidence of oil spills resulting from tanker traffic, offshore drilling, and associated activities, will increase in years to come as the world's demand for petroleum and petroleum products continues to be on the rise.

The fate of an oil spill in the marine environment is determined by the apparently complex and interrelated processes of evaporation, dissolution, photochemical and microbial degradation, emulsification, sedimentation, and sinking. The physical and chemical alterations to the spill occurring with time, as well as the rates of these changes, will be influenced by a variety of abiotic environmental parameters, as well as by the physical-chemical properties inherent to the oil itself. There are large gaps in our knowledge with respect to the understanding of the "mass balance" involved in the environmental degradation of an oil spill. The major processes related to the fate of petroleum spills in the marine environment are depicted in the figure on the following page from Burwood and Speers (36).

A number of excellent reviews have been published concerning the fate and effects of petroleum in the marine environment, including Clark and Brown (47), Clark and MacLeod (48), Karrick (120), the National Academy of Sciences (171), McAuliffe (156), Wheeler (237), and others. In this report some of the works cited in these previous reviews are touched upon; however, we have tried to concentrate, for the most part, on the more recent publications dealing with oil weathering, which have appeared in the literature since 1977. Where there have been omissions or gaps in existing literature, the authors can only apologize for the oversight.

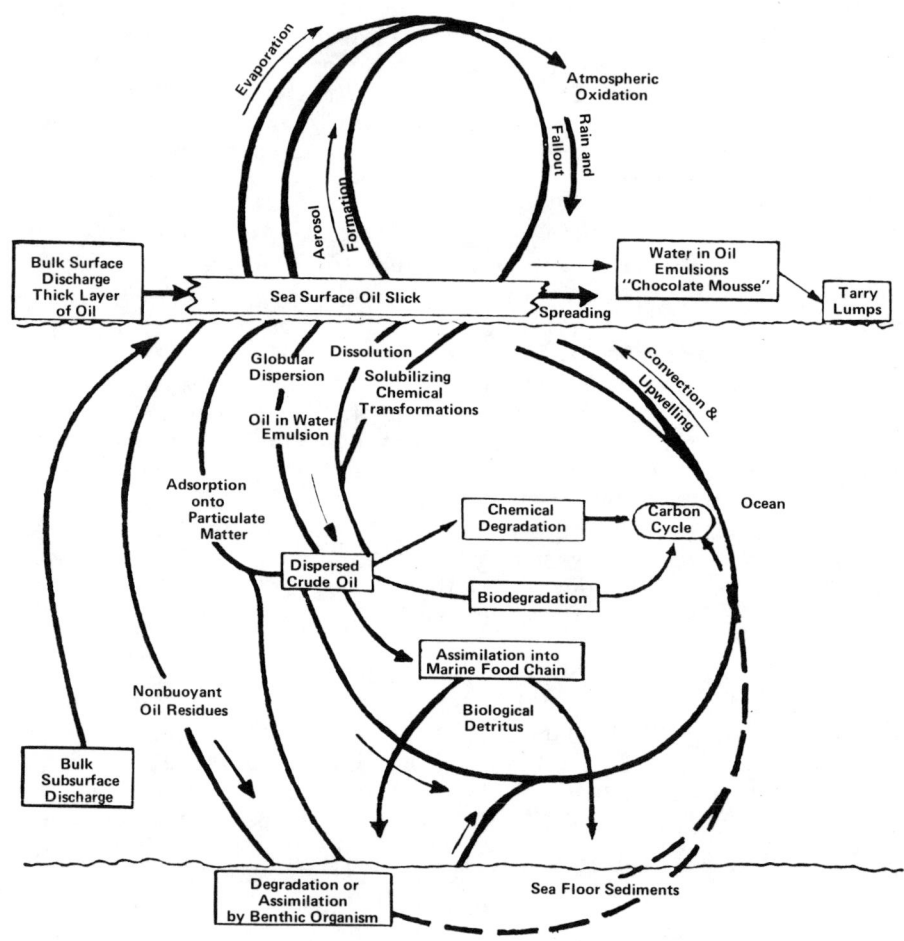

A schematic overview of the various combined and competing weathering processes that act on spilled oil in the marine environment (from Burwood and Speers, 36). Reprinted with permission from Estuarine and Coastal Marine Science, Vol. 2, © 1974 by Academic Press, Inc.

# I. ABIOTIC FACTORS/PROCESSES

This section deals with the physical and chemical changes occurring in a petroleum spill with time as a result of interactions with environmental parameters and system fluxes. Attention will be given to not only how petroleum composition influences, and is influenced by, a particular abiotic process, but also to composite influences derived from the interrelations of these processes.

## SPREADING

Spreading of oil released on water is probably the most significant process for the first six to ten hours following a spill; Figure I-1 presents the four main forces acting on spill during this process (78, 237). Figure I-2, from R. B. Wheeler (237), illustrates that in addition to spreading there are other competing forces which occur simultaneously once oil has been released onto the water.

The principal forces influencing the lateral spreading of oil on a calm sea are gravitational (causing decreasing film thickness), surface tension, inertial forces, and frictional forces (237). The gravitational spreading force is proportional to the film thickness, the thickness gradient, and the density difference between the water and the oil. The second force causing spreading is denoted by the spreading coefficient, sigma, which is the difference between the air/water surface tension and the sum of the air/oil and oil/water surface tensions. This force is independent of film thickness, and it becomes the dominant process in the final phases of spreading. Only aromatic and aliphatic hydrocarbons with less than nine carbons (171) have positive spreading coefficients, so compounds with greater molecular weights do not tend to spread on water (171). Fay (78) stated that spreading ends when the removal of volatile fractions reduces the spreading coefficient to zero.

The spreading forces due to gravity and surface tension are retarded by the inertia of the oil body and by oil/water friction. The inertia of a specific oil slick is a function of the thickness and the density, and it diminishes rapidly as spreading proceeds. Fay has reported that inertia is probably only important for the first 1,000 to 10,000 seconds after a spill (78). Concurrent with the spreading process, friction between the slick and the surface water increases as the thickness of the oil slick itself decreases and the viscous water/oil mixture increases. This frictional retardation force eventually overcomes the inertial forces, and it remains the dominant anti-spreading force.

# 4  FATE AND WEATHERING OF PETROLEUM SPILLS

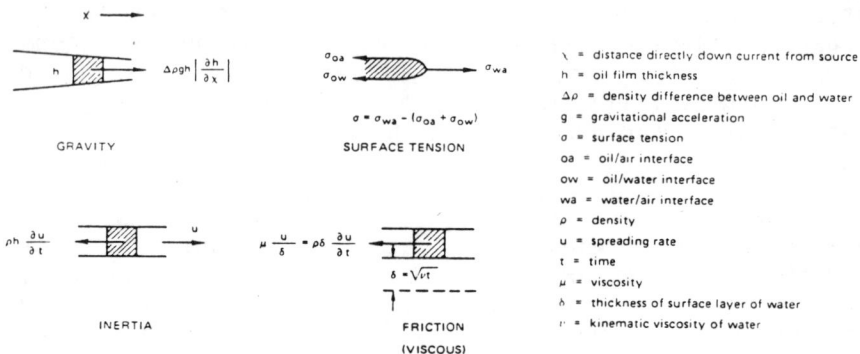

Figure I-1.  The four forces that act on an oil film (from Fay (78) as cited by Wheeler, 237).  Reprinted with permission from the author and the Exxon Production Research Co.

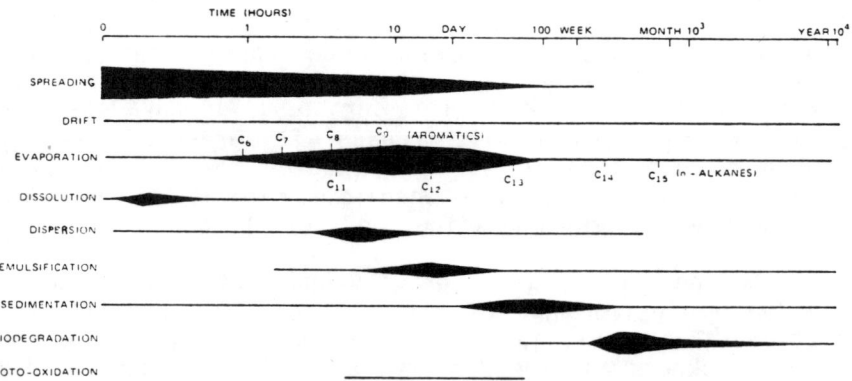

Figure I-2.  Process vs time elapsed since the spill.  The line length indicates probable timespan of any process.  The line width indicates the relative magnitude of the process through time and in relation to other contemporary processes (from Wheeler, 237).  Reprinted with permission from the author an the Exxon Production Research Co.

Buckmaster (35) has derived an equation which describes the gravity viscous spreading phase of oil; however, the mathematical model only holds for spreading on calm seas which is the exceptional case. In the open sea, turbulent wind/water activity dominates the distortion and distribution of the spreading slick and, in general, the most important factors are the current and wind vectors which direct the slick. With regard to spreading, however, Stolzenback et al. (213) stated that spreading is primarily controlled by short time period (less than $10^3$ seconds) and small scale (less than $10^3$ meters) atmospheric fluctuations. Longer time scale atmospheric perturbations become important when considering "Drift", as discussed in the following section. Murray (170) has proposed the application of different diffusion theories for the prediction of spreading, and suggested that when comparing the surface tension theory of Fay (77) and Taylor diffusion theory, that Taylor diffusion theory more accurately represents slick behavior (170). Brief discussions of Fay's work and several comparisons of Taylor theory versus Fickian diffusion theory are presented by Wheeler (237).

As the spreading process occurs, the oil is more rapidly weathered and this brings on additional changes (Figure I-2). Within several hours, over 90% of the hydrocarbons lighter than $n-C_{10}$ are removed by evaporation and dissolution, and this leads to a decrease in the slick volume and an increase in the viscosity and specific gravity, as discussed later. Inhomogenieties in the thickness of the slick can also occur as a result of differential spreading rates among different hydrocarbon components. In the Argo Merchant spill, for example, oil pancakes up to 15 cm thick with surface areas of from $10^2$ to $10^4$ meters$^2$ were reported (172). This difference in spreading coefficient may be related to molecular weight, as Phillips and Groseva (187) experimentally determined that the spreading coefficient varied inversely with molecular weight.

Water temperature is also a factor in the spreading; however, it generally does not affect most oils which have a viscosity and density less than that of Bunker C.

The spreading coefficient has also been found to be proportional to the NSO content of the oil (84) and once the water-soluble NSO compounds, which are more polar than the residual hydrocarbons, are dissolved, this spreading tends to stop (78, 117).

## DRIFT

Drift is a large scale phenomena and is a measure of the movement of the center of mass of an oil slick. Drift is primarily controlled by wind, waves, and surface currents and is independent of the spreading and spill volume. For example, when wind speed is the dominant force in drift movement, it has been demonstrated that a slick can move at a rate of up to 3% of the wind speed (173, 208). However, prediction of slick drift due to wind patterns alone is difficult, because of wind and wave perturbations. This is compounded by the fact that most surface wave spectra are composed of a number of different wave systems with different periods and directions. Furthermore, most models use wave functions generated by local winds over a specific time period, and in real ocean situations the components of the wave spectrum are swell and waves which often reflect wind generated seas at a time and locality remote from the slick (237). Additional modeling complications result from wave test tank data which demonstrate that, in addition to waves affecting oil slicks, oil slicks also tend to have a concomitant impact upon the resultant wave spectra. In lab tests using diesel oil Number 2, Liu and Lin (137) demonstrated that oil on water significantly decreased the amplitude of wind waves with increasing wind speeds of up to 4-10 m/sec. Wind speeds above 10 meters/sec broke the oil slick up into smaller lenses (137).

Nevertheless, wind direction and duration can have a dramatic effect on oil spill drift and trajectory, and an example of wind induced drift perturbations of an extensive near shore spill occurred after the wreck of the Amoco Cadiz. After that spill, the windshift pattern which occurred on April 2, 1978, caused the extensive oiling of previously clean coastal areas which had not been touched during the first eight to ten days after the spill (104). Figure I-3 shows the wind direction changes which occurred in the month following the wreck.

Drift is also strongly influenced by waves and tidal currents. After the Amoco Cadiz spill, waves of 1.5-2 meters and 6-7 meter tides further affected the overall drift and the extent of the oiling of the area (104). At the end of the first two weeks a total of 72 kilometers of coastline had been covered with oil, then, following the dramatic wind shift change of April 2, a total of 213 kilometers were lightly oiled and 107 kilometers were heavily oiled (104).

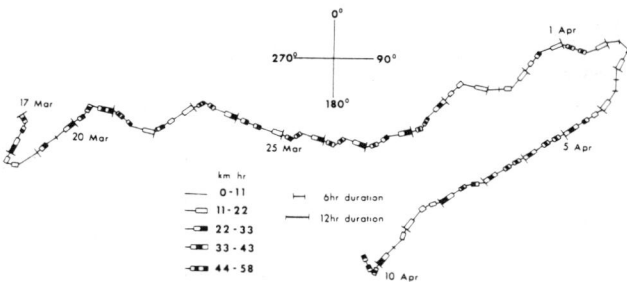

Figure I-3. Wind pattern for March 17-April 10 from the French meteorological station 1 km north of l'Aber Wrac'h. The wind shift on April 2 caused the oiling of previously clean coastal areas south of the wreck site (from Hayes, 104). Reprinted with permission from the author and the Oils Spill Conference Office, Washington, DC.


Venkatish et al. (226) have developed open-sea ice free models for oil spill (drift) movement, and they have partially verified the results with simulation experiments conducted in the Bay of Fundi. With observed winds, the model predicted locations of oil on water (as estimated with "Orion" buoys) fairly well, when compared to the buoys actual locations. It was very clear from their work, however, that what is needed for better modeling of drift are high resolution surface wind forecasts. In their work (226), the oil was treated as independent aggregates of parcels, and then a detailed computer model and an interactive set of equations were used to compute the alterations. Among the parameters considered were evaporation, emulsification under different sea states, surface water current movement (advection) and random motion due to turbulence. Wind predictions proved to be the weakest link in this program. Figure I-4 presents the predicted and observed location of the Orion buoys with forecast winds, as opposed to Figure I-5, which presents the observed and predicted location of the Orion buoys with observed and (actual) winds measured at the site (i.e. an after the fact evaluation or computer simulation of the buoy release). Clearly, better determination of wind field vectors is critical for the monitoring of slick advection. For this consideration, only patterns on a scale of weather systems--with fluctuation periods of greater than $10^5$ seconds or a lateral extent of $10^5$ meters are really significant (237). As noted earlier, smaller scale fluctuations are primarily important only for spreading, as opposed to drift mechanisms. Obviously, in near-shore spills, the effects of spreading as well as advective drift will be important in trying to prevent contamination of shorelines.

8  FATE AND WEATHERING OF PETROLEUM SPILLS

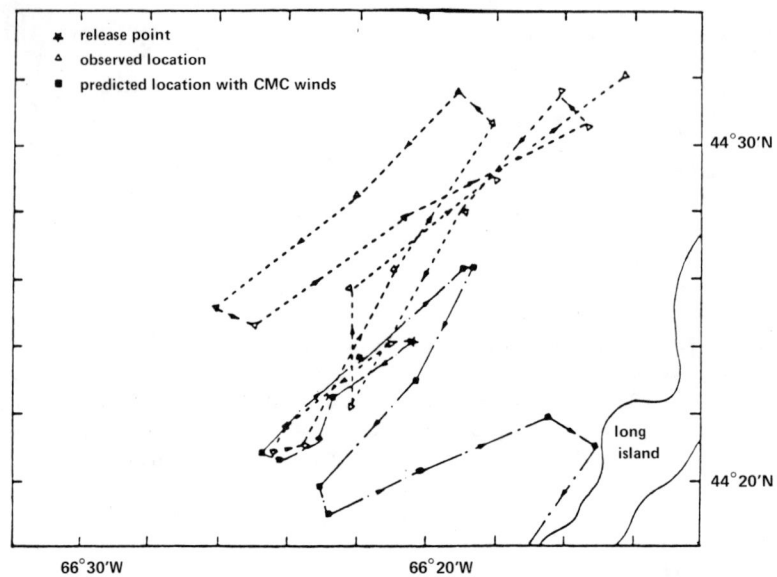

Figure I-4. Predicted and observed location of the Orion buoys with forecast winds (from 1979 Oil Spill Conference, S. Venkatesh et al., 226). Reprinted with permission from author(s) and the Oil Spill Conference, Washington, DC.

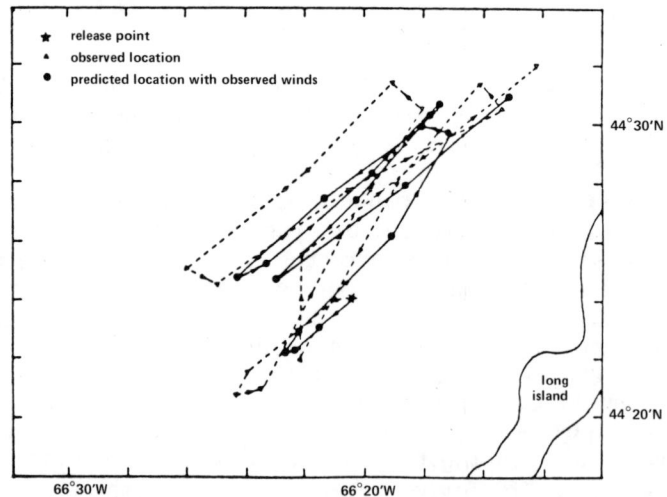

Figure I-5. Observed and predicted location of the Orion buoys with observed (actual) winds measured at the site (from 1979 Oil Spill Conference, S. Venkatesh et al., 226). Reprinted with permission from author(s) and the Oil Spill Conference Offices, Washington, DC.

## SPREADING AS INFLUENCED BY DETERGENTS OR EMULSIFYING AGENTS

According to Garrett (84), small quantities of polar organic compounds are sufficient to modify the behavior of oil in water. For example, in tests using 1-dodecanol and a paraffin oil with a spreading coefficient of -11.7 dynes/centimeter, 0.2, 0.5 and 1.0% 1-dodecanol dissolved in the oil altered the spreading coefficients to values of 3.0, 11.0, and 17.5 dynes/cm, respectively. The orientation of the surface active molecules lowers the interfacial tension and thus modifies the viscosity differences at the oil/water boundary leading to enhanced spreading and formations of emulsions when the oil layer is agitated by wind or waves (84). Thus, crude petroleum and most types of refined products which spread across the surface of relatively quiet or confined waters can be cleaned up or dispersed with these polar compounds. No satisfactory method has been found, however, for dealing with heavy fuel oils that tend to solidify when spilled into cold water (72). Spreading is also affected very much by surface active constituents which are important in enhancing the dispersion of oil in water. For the most part, spreading is enhanced by increasing the number of NSO type compounds present in the mixture. The composition of various dispersants varies; however, they are generally comprised of a petroleum hydrocarbon base to enhance oil solubility, along with a solvent containing some aromatics and saturates. BRAS-X-PLUS, a dispersant used in the cleanup of the Brasilian Marina incident (62), has a petroleum hydrocarbon base of 75% with 25% of this solvent being aromatics, including naphthalene and alkylbenzenes, and 75% being saturates and cyclohexane. The boiling point is similar to a low-end kerosene or jet fuel (450-500$^\circ$F) to a gas oil on the high end. After evaporation of the detergent in a steam bath which removed 72% of the material, the 28% residue was primarily gas oil and surfactant. This surfactant contained ethanol and amide compounds which obviously acted as the surface active components.

Chemical detergents which are often used as emulsifying agents tend to increase the formation of oil in water emulsions and thus facilitate the dispersion of oil by stabilizing these emulsions when formed. In a 1970, 34$^\circ$ API gravity petroleum blowout in the Gulf of Mexico, detergents (approximately 3.5% of the quantity of oil discharged) were used in the water sprayed on the burning platform and the resultant plume was intensively emulsified and dispersed to undetectable levels at relatively short distances from the platform (154). The emulsifier caused the rapid downward mixing of the oil and under the spill conditions, the emulsified oil was observed only 1-1.5 miles from the spill site (by chemical analysis). The untreated oil could be observed 6-9 miles in a plume downstream of the site. Also, it was noted that the size of the oil/emulsifier particulates was dependent on the amount of emulsifier added (156). Once mixed with the detergent and clear of the platform, the oil developed three major modes of appearance. The most noticeable was a bright, reddish-brown band of thick oil, roughly three meters in width, extending for many kilometers on the sea surface. This appeared to be a sea-water-in-oil emulsion which was presumably induced by the detergents or the high pressure water streams used to quench the fire, or both. The second mode of appearance was a thin, dull grey through an irridescent to silverish sheen which became wider as a surface plume or slick down in from the platform. The third mode was a creamy, yellow subsurface plume emanating from the platform. This was believed to be an oil-in-seawater emulsion with the oil dispersed into very fine droplets by the action of the detergent (2).

Somewhat different results were obtained in spilled oil tests completed by Buckley and Humphrey (34), who found that in spite of the lack of a measurable density gradient (up to 30 m in depth) in a controlled spill situation, there was no significant diffusion of dispersed oil deep into the water column over time when treated with the dispersant Corexit 9527. The integrated quantity of oil in the water column decreased more quickly than did either the mean or maximum oil concentration of the horizontal cloud, indicating that some of the dispersed oil may have been rising back to the surface. Further, they observed that the dispersed oil concentrations measured in their controlled spill situation were lower than anticipated. This was due to the slick breaking up into numerous smaller slicks and patches. Therefore, the dispersion was not as efficient as anticipated, in that some small patches were missed. They also found that dispersion of oil into the water column retarded, or inhibited weathering. Further, the dispersed cloud of oil moved with the surface layer of the water, and while it spread more rapidly than predicted by models, the rate of spreading did not increase with time, as predicted for open ocean situations. These differences are likely due to the location of the experiment within a sheltered cove. As noted above, there were also some indications that the chemically dispersed oil did not stay mixed with the detergent in the water column, but came slowly back to the surface. Thus, although the use of dispersants as described above has had mixed success in enhancing some open ocean or near-shore spilled cleanup efforts, its indiscriminate use should be avoided in that oil dispersant mixtures have been shown to have severe ecological effects in different environments.

In fresh water, Scott et al. (202) showed a dramatic drop in the dissolved oxygen concentrations with time as a result of a dispersant. Figure I-6 shows the decrease in dissolved oxygen concentration as a function of time. It was also noted in these studies that the dissolved organic carbon and soluble silica content dramatically increased, as a result of dispersant or detergent treatment.

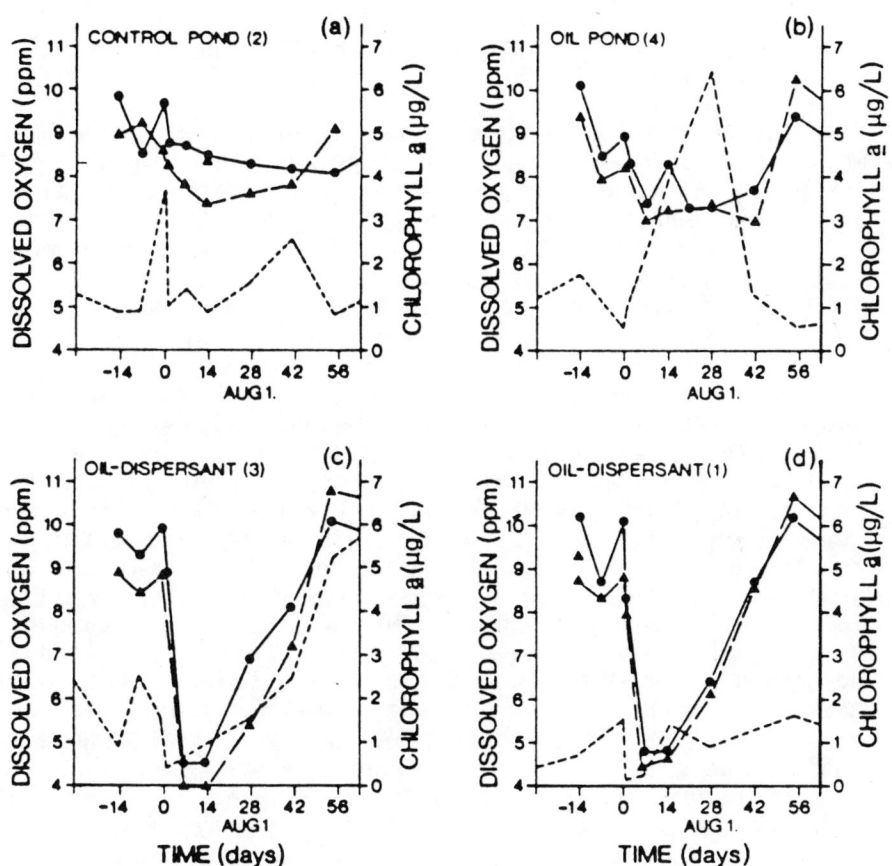

Figure I-6. Decrease in dissolved oxygen concentrations as a function of time, shown in solid lines, where ● indicates subsurface levels and ▲ indicates near-bottom levels. Chlorophyll-a values are plotted against time by a broken line (from Scott et al., 202). Reprinted with permission from author and the Oil Spill Conference Office, Washington, DC.

## EVAPORATION

Evaporation and dissolution are the two major processes of degradation of oil after it is released into the water. Figure I-2 represents the process of evaporative loss as a function of time and illustrates that in addition to the other degradive processes acting on the oil, the majority of the more volatile compounds may be lost by evaporative processes within 24-28 hours. The composition of the oil, its surface area and physical properties, the wind velocity, air and sea temperatures, sea state and the intensity of solar radiation, all affect hydrocarbon evaporation rates (237).

Evaporation removes most of the volatile lower molecular weight compounds and, in general, pentadecane ($n-C_{15}$) is the lowest normal alkane commonly found in weathered oils; also, presumably due to evaporation, it is rare to find hydrocarbons lower than $n-C_{12}$ in seawater extracts. These more volatile components make up 20-50% of most crude oils, 75% or more of refined petroleum fuel and 10% or less of residual fuel oils, such as Bunker C (39).

When a water-in-oil dispersion of Number 2 fuel oil was added to the Marine Ecosystems Research Laboratory (MERL) test tanks for the determination of the fate of petroleum in a marine ecosystem, the primary loss of the hydrocarbons was due to evaporation, with 60-90% of the total material being removed in this manner (85). Preliminary tests on the Ekofisk Bravo oil slick indicated that over 60% of the original components had evaporated by the ninth day (31). In a detailed study of the chemical weathering and composition of oil in samples of mousse (water-in-oil emulsions) collected after the Amoco Cadiz spill, Calder (41) noted that evaporation was an important weathering process for all of the low boiling components of the oil, including aliphatic and aromatic hydrocarbons and even heteroatomic compounds such as benzothiophene. Figure I-7 shows the decrease in the relative composition of hydrocarbons by boiling point from the parent material, and shows that a three to four-fold loss in materials boiling below $225^{o}$ occurred within 2-3 days (41). Weathering rates of selected hydrocarbon losses (including both evaporation and dissolution) are shown in Figure I-8, indicating that greater than 40% of the compounds with molecular weights less than $n-C_{12}$ were lost per day and that residence times of from 2-5 days could be calculated for compounds below $n-C_{15}$ (41). It is interesting to note that Calder also found substantial

# ABIOTIC FACTORS/PROCESSES 13

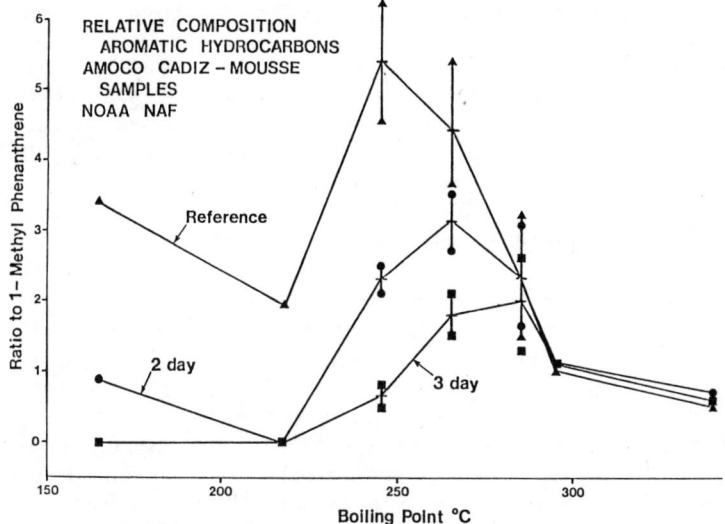

Figure I-7. Decrease in the relative composition of hydrocarbons and heteroatomic compounds (from Calter, 41). Reprinted with permission of author and NOAA/OCSEAP Office, Boulder, CO.

| Compound | % loss/day (rel.) | Residence time (days) in mousse (calc.) |
|---|---|---|
| TMB, N, <nC 12 | >40 | 2 |
| $C_1$-N, nC 13, nC 14 | 30 | 3 |
| $C_2$-N, nC 15−nC 18 | 20 | 5 |
| nC 19−nC 23 | 10 | 10 |
| $C_3$-N, nC 24−nC 28 | 5 | 20 |
| Fl, Ph, >nC 28 | 0 | ∞ |

Figure I-8. Weathering rates of hydrocarbons from mousse (from Calder, 41). Reprinted with permission of author and NOAA/OCSEAP Office, Boulder, CO.

concentrations of volatile aromatics in some sediment samples as well as in selected mousse and water samples, indicating that in several instances some factor had depressed the rate of evaporation. At the time of collection, the mousse from some of these samples was several inches thick. If similar thicknesses existed in other samples as they moved from the Amoco Cadiz, a lower rate of volatilization could be expected, due to the limited migration of the volatile components to the surface of the mousse.

In early samples taken after the Amoco Cadiz spill, the $n-C_{17}$/pristane and $n-C_{18}$/phytane ratios were still relatively constant compared to the parent oil samples, indicating that bacterial activity had not yet commenced. In the parent oil, the most predominant n-alkane was $n-C_{11}$. In samples collected that were assumed to be approximately half a day oil, major losses, presumably due to volatilization, in the $n-C_{10}$ to $n-C_{12}$ region occurred as determined by gas chromatographic analyses. The highest concentration of an alkane component in this sample was $n-C_{13}$. Alkyl substituted aromatic compounds in the same boiling point range were also lost due to evaporation and/or dissolution with methyl-naphthalenes becoming less abundant than the dimethylhomologues. In samples that were three days old, trimethylnaphthalenes were more abundant over other naphthalene isomers (41). During this time period the higher molecular weight phenanthrenes and dibenzothiophenes did not diminish. After three to five days, most of the highly volatile components were removed from the mousse.

When oil is spilled in an Arctic environment or on ice covered waters, evaporative processes continue to occur, although at a lower rate. Following the spill of the Bouchard #65 in Buzzard's Bay during the winter of 1977, evaporative weathering was directly dependent on the field conditions which determined the extent of exposure (19). Table I-1 shows the percent loss due to evaporation of the Number 2 heating oil as a function of the field condition. The data for these analyses were generated by glass capillary gas chromatography using the oil from the cargo hold as a reference. Approximately 85% of the reference cargo oil consisted of saturated aliphatics, ranging from $n-C_9$ to $n-C_{23}$. The remaining 15% of the cargo was primarily aromatic hydrocarbons, ranging from one to three substituted aromatic rings.

The results in the table indicate little evaporative loss (less than 10%) in 7 of the 11 spilled oil samples. The least weathering occurred in 2 samples which were obtained from under the ice where less than 4% of the material was evaporated. It appears that the four samples showing the

Table I-1. Percent Loss caused by evaporation of No. 2 heating oil as a funcbion of the field condition (from MESA report on Bouchard #65 Oil Spill, Baxter et al., 19). Reprinted with permission of NOAA/Science Applications, Inc. MESA Program, Boulder, CO.

| Location | Field Condition | Percent Loss Alkanes (85% of cargo) | Approximate Percent Loss Arenes (15% of cargo) | Approximate Total Percent Loss |
|---|---|---|---|---|
| Wings Neck Tower | Oil underneath ice near edge of rafted ice. Oil was approx. 1.3 cm thick | 4 | 14 | 6 |
| Wings Neck Tower | Oil in ice sheltered by overlying ice sheet | 4 | 16 | 6 |
| Wings Neck Cove | Oil in ice near edge of ice floe. Sample taken from top 38 mm of ice core. Medium stained ice (Section 3.2.3 defines light, medium and heavily stained oiled ice) | 5 | 20 | 8 |
| Wings Neck Cove | Slush oil/snow mixture from shallow oil pool in hummock | 7 | 31 | 12 |
| Wings Neck Cove | Oil taken from rafted oil pool | 7 | 38 | 13 |
| Wings Neck Cove | Heavily oil stained ice from ice floe near edge of small oil pool | 10 | 33 | 15 |
| Wings Neck Tower | Ice piece 0.3 mm thick taken from small pressure ridge. Ice appeared to be medium stained | 9 | 38 | 15 |
| Wings Neck Tower | Wind blown oil on top of ice | 21 | 54 | 28 |
| Wings Neck Tower | Wind blown oil on top of ice | 25 | 64 | 33 |
| Wings Neck Cove | Ice piece rotated in air. Scraped off top of medium stained oily ice | 29 | 58 | 35 |
| Wings Neck Cove | Ice piece rotated in air. Scraped off top of lightly stained oily ice | 37 | 89 | 47 |

greatest losses, in the range of 20-30%, came from the more exposed samples. The aromatics appeared to be preferentially lost by volatilization (14-89%), whereas the aliphatics were lost in the range of 4-37%. In both cases, these losses clearly depended on the conditions of exposure, and almost exclusively, the evaporative losses were restricted to compounds having molecular weights less than $n\text{-}C_{16}$.

In attempting to determine a theoretical basis for estimating evaporation rates of hydrocarbons from a mixed hydrocarbon substrate, Fallah and Stark (75) proposed the following equation:

$$\frac{dv}{dt} = KA^\beta \, U(z)^\alpha \, (P_s - P_a)$$

Where A is the liquid surface area; $U(z)$ is the wind speed at a height z above the liquid surface, $P_s$ is the saturation vapor pressure at the liquid surface temperature, $P_a$ is the partial vapor pressure in the air upwind of the liquid surface, V is the volume, t is time, and K, $\alpha$ and $\beta$ are constants.

The evaporation rate for specific hydrocarbons is a function of its vapor pressure, which in turn is inversely related to the molecular weight. Figure I-9 presents the vapor pressure as a function of carbon number for alkanes, isoalkanes, cycloalkanes, and aromatics (198, 237). From this figure it would appear that aromatics may tend to lag slightly behind the paraffins in evaporation. However, due to the greater solubility of aromatics, their overall loss from a slick due to both volatilization and dissolution may be higher.

Compounds with molecular weights greater than $n\text{-}C_{15}$ will continue to evaporate over extremely long period of time although these rates become insignificant after $10^2$ hours. Thus, compounds with vapor pressures greater than that of $n\text{-}C_8$ will not persist in a slick, whereas, those with vapor pressures less than $n\text{-}C_{18}$ do not evaporate appreciably under normal conditions (192). As mentioned above, these characteristics suggest that in many cases up to 50% of an oil spill may be evaporated within the first 24 hours. Spills of refined products, such as kerosene and gasoline, which have primarily lower boiling components, may be completely lost by evaporative processes.

ABIOTIC FACTORS/PROCESSES   17

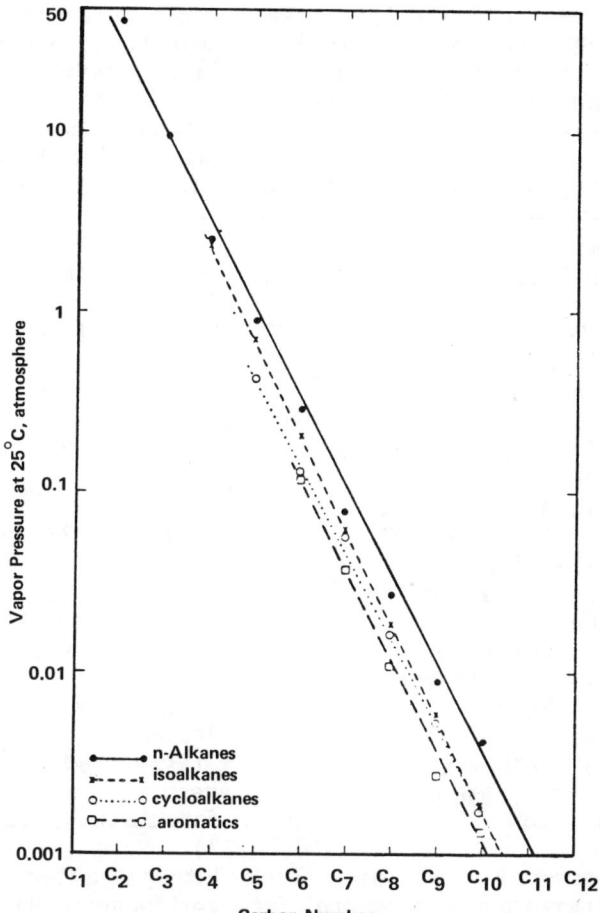

Figure I-9. Vapor pressure as a function of carbon number for alkanes, isoalkanes, cycloalkanes and aromatics (from Wheeler, 237). Reprinted with permission from the Exxon Production Research Co.

The sea state can also affect the evaporative loss, and Table I-2 presents data from Rostad (199) which illustrates the evaporative losses as a function of the Beaufort sea state. In low sea states, evaporative loss was estimated to be 30% during the first day, and it went as high as 40% during the high Beaufort sea state, stage 6. Specifically, evaporation in sea states greater than 5 is increased due to the increased wind speed and increased surface area of the oil exposed to the atmosphere, thereby enhancing molecular exchange, and due to the formation of aerosols due to spray and breaking waves. In extremely rough sea states the weathering is thus enhanced, and Smith and MacIntyre (207) found a sharp increase in the evaporation of n-paraffins

Table I-2. Evaporative losses as a function of the Beaufort sea state (From Rostad, 199). Reprinted with permission of the author and the Royal Norwegian Council for Scientific and Industrial Research.

| Sea States (Beaufort Scale) | Low (2-3) | Med. (4-5) | High (6) | No Dispersant Action (7) |
|---|---|---|---|---|
| Spill volume, $m^3$ | 100 | 100 | 100 | 100 |
| Evaporation loss I (first day), % | 30 | 35 | 40 | 50 |
| Residual oil volume A, $m^3$ | 70 | 65 | 60 | 50 |
| Primary cleanup efficiency, % | 70 | 50 | 30 | |
| Residual oil volume B, $m^3$ | 21 | 32.5 | 42 | 50 |
| Evaporation loss II (5·days), % | 20 | 15 | 10 | |
| Residual oil volume C, $m^3$ | 16.8 | 27.6 | 37.8 | 50 |
| Natural dispersion--daily loss rates for 5 days only, % | 10 | 15 | 25 | 30 |
| Residual oil volume D before second cleanup, $m^3$ | 9.9 | 12.3 | 9.0 | 8.4 |
| Secondary cleanup efficiency, % | 40 | 30 | 20 | |
| Residual oil volume E reaching the beaches, $m^3$ | 5.9 | 8.6 | 7.2 | 8.4 |

from a Number 2 fuel oil slick at sea after the wind velocity increased enough to form whitecaps and breaking waves. In general, petroleum hydrocarbons which are removed from the sea surface as aerosol particles do not remain in the atmosphere. Usually, the particles are redistributed back into the marine environment in a matter of seconds for particles larger than 0.1 millimeters, and up to a few days for micrometer sized particles (2). Baier (12) suggested that the bursting of bubbles in the formation of spray could explain the removal of thin oil films or sheens over time.

In a variety of field observations on the loss of volatile hydrocarbons from sea surface, it was noted that the rates of loss of volatile hydrocarbons under various environmental conditions on the open ocean were faster than those observed in laboratory studies. Table I-3 summarizes the data from five field observations and two laboratory studies (156). In the laboratory study, Regner and Scott (192) found that 10% of the $n$-$C_{12}$ and lower hydrocarbons remained in a fuel oil sample after 80 hours of exposure at $20°C$ with air movement of 11 knots.

Table I-3. Loss of volatile hydrocarbons from oils on water surface (from McAuliffe, 156). Reprinted with permission of Pergamon Press Ltd., copyright 1977.

| | Open Seawater | | | | | Laboratory | |
|---|---|---|---|---|---|---|---|
| | Harrison et al. (1975) | Kinney et al. (1964) | McAuliffe (1977) | Sivadier and Mikolaj (1973) | Smith and MacIntyre (1973) | Kreider (1971) | Regnier and Scott (1975) |
| Type of Oil | Crude | Crude | Crude | Crude | No.2 Fuel | Crude | No. 2 Fuel |
| No. of Spills | 5 | 1 | 4 | 2 | 1 | Several | Several |
| Vol. Spills, bbl | 6.6 | Small | 10.5 | 0.02 | 4.8 | 450 ml[a] | |
| Water Temp., °C | 24 | 25 | 11-14 | 19-20 | 5 | | 20 |
| Air Temp., °C | 21-24 | | 12-17 | 28-30 | 5 | | 20 |
| Wind Speed, knots | 0-18 | 9-12 | 8-24 | 8-12 | 1-18 | | 11.3 |
| Waves | Calm to Whitecaps | | 1-5 ft | | Calm to Whitecaps | | Wavelets |
| Time for Hydro-Carbon Loss | | | | | | | |
| $C_9$ and lower | 40-90 min | 8 hr | 4-8 hr | 90 min? | 7 hr | 24 hr | 80 hr (10% left) |
| $C_{10}$ and lower | | | | | | | |
| $C_{12}$ and lower | 3-8 hr | | | | | | |

[a] Oil was added onto water in a 55-gal drum outside laboratory in Richmond, CA.

Harrison et al. (103), on the other hand, found loss of n-$C_{12}$ to be complete in three to eight hours after a spill from the sea surface. Figure I-10 shows the percent of low boiling aromatic hydrocarbons remaining in a South Louisiana crude oil spill as a function of time (156). The rapid decrease after 42 minutes of n-$C_{12}$ and n-$C_{13}$ represents the onset of strong winds causing whitecapping. McAuliffe (157) reported similar losses for the aromatic compounds ranging from benzene to trimethylbenzenes as shown in Figure I-11. From these results it is apparent that the more volatile and toxic hydrocarbons such as benzene and toluene are rapidly removed from an oil slick by evaporative processes. These compounds are also the most soluble in water and thus their removal could be enhanced by dissolution as well. To test the relative importance of evaporation and dissolution in the removal of hydrocarbons from slicks, concentrations of benzene and cyclohexane were compared (103). Benzene and cyclohexane have similar vapor pressures: 95.5 mmHg for benzene and 97.8 mmHg for cyclohexane. Therefore, these compounds should evaporate with similar rates. However, their solubilities are quite different, (1,789 µg/l for benzene, 55 µg/l for cyclohexane) suggesting that benzene might be preferentially removed by dissolution. Measured concentrations remaining in the oil after several spills, however, showed no preferential removal of either benzene or cyclohexane (103). This suggests that evaporative processes dominate over dissolution in the removal of volatile components from the slick. From these experiments it was predicted that losses of aromatic hydrocarbons from evaporation would be 100 times faster than losses from dissolution and that the evaporative rate for aliphatics may be 10,000 times greater than that for dissolution (103).

Smith and McIntyre (207) have also found that the total loss of components from three refined products, Number 2, Number 4, Number 6 (Bunker C) fuel oils from dissolution in artificial sea water was negligible compared to that lost by evaporation. In general, the dissolution was determined to be 0.2 to 1.1 percent of evaporation (207). Dissolution thus appears to be a degradative process of significant magnitude only within the first several hours after a spill. Although dissolution can continue for several hours, the process is believed to be relatively insignificant compared to that of evaporation.

Changes in the physical properties of oil resulting in the loss of volatile hydrocarbons include increases in density and the kinematic viscosity as shown in Table I-4. These alterations inhibit spreading and molecular diffusion of the

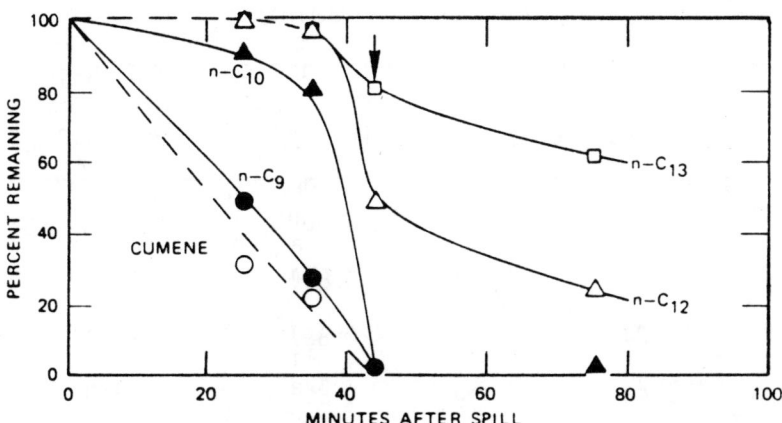

Figure I-10. Percent of low-boiling hydrocarbons remaining in south Louisiana crude oil slick (from McAuliffe, 156; adapted from Harrison et al., 103). Reprinted with permission of Pergamon Press Ltd., © 1977.

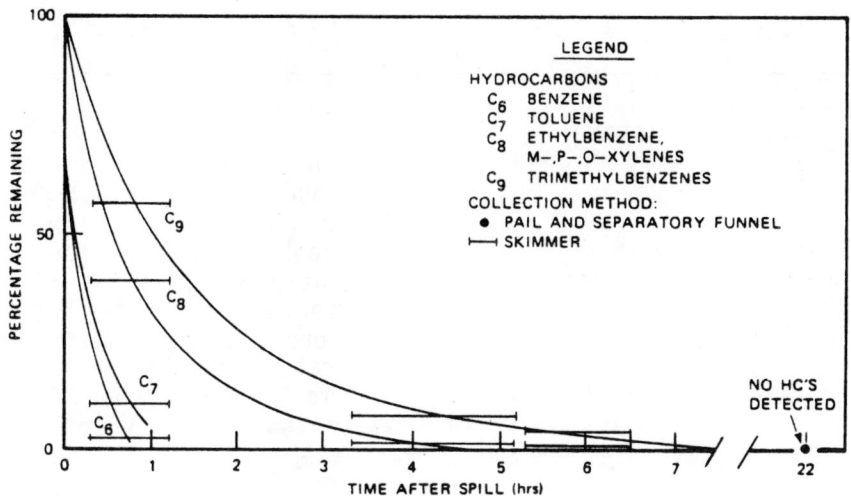

Figure I-11. Percent of aromatic hydrocarbons remaining in surface oil slick--first La Rosa spill (from McAuliffe, 157). Reprinted with permission from Pergamon Press, Ltd., © 1977.

Table I-4. Density and kinematic viscosity of two weathered crude oils (from Rostad, 199). Reprinted with permission of the author and the Royal Norwegian Council for Scientific and Industrial Research.

| Oil | Time Weathered, hours | Density, g/ml | Kinematic Viscosity, $mm^2/sec$ [a] |
|---|---|---|---|
| A | 0 | 0.862 | 18.78 |
| A | 8 | 0.890 | 40.13 |
| A | 24 | 0.891 | 47.07 |
| A | 48 | 0.894 | 56.02 |
| A | 168 | 0.903 | 97.53 |
| B | 0 | 0.851 | 15.40 |
| B | 8 | 0.896 | 28.39 |
| B | 24 | 0.901 | 34.29 |
| B | 48 | 0.908 | 53.88 |
| B | 168 | 0.914 | 73.96 |

[a] $mm^2/sec$ = centistoke.

remaining oil components, and in this regard, the evaporation of petroleum hydrocarbons eventually becomes self-limiting. As the more volatile hydrocarbons are lost, the viscosity of the resulting oil increases and this can lead to the breakup of slicks or patches of oil into smaller droplets. Agitation of these droplets enhances incorporation of water due to increased surface area, and these smaller particles have been observed to aggregate into thicker bodies of oil (J. R. Payne, personal observation, IXTOC I cruise). Diffusion of the more volatile components within these bodies then becomes slower and can become the limiting factor in the loss of the remaining volatile components (156).

## PHOTOOXIDATION

### Introduction

In the presence of oxygen, natural sunlight has sufficient energy to transform many petroleum hydrocarbons into compounds possessing significant chemical and biological activities. The mechanism of this photo-involved process can be described as an autocatalytic free-radical chain reaction (36, 82, 128, 142), which results in formation of hydroxy compounds, aldehydes, ketones, and ultimately, low molecular weight carboxylic acids. Higher molecular weight intermediates can be formed concomitantly, via radical recombination (polymerization) or condensation reactions (82) of aldehydes and ketones by phenols, or by esterification between alcohols and carboxylic acids (184). Sunlight may also cause copolymerizations by way of thermally induced oxidations, such as the copolymerization with oxygen by methylstyrenes and indene (109).

Formation of intermediate hydroperoxides has been substantiated (36, 128, 129), and hydrocarbons suitable for photo-initiated hydroperoxide formation are widely represented in crude oils as isopropyl-substituted aromatics and alkyl-substituted aracyclane and cyclane structures (36).

A kinetic study on photochemical weathering (142) attempted to fit spectrophotometric (UV) data with various kinetic models, and suggested the most probable as being a second-order autocatalytic process. The proposed mechanism involved reaction of oxygen with reactant to produce an intermediate radical, which in turn results in formation of an oxidation product capable of forming additional free radicals itself. These radicals further react with reactant and product molecules to produce more oxygenated products, and so on.

Another study (128) provides credence to this mechanism with respect to the conversion of fluorene. This mechanism proposed the conversion of fluorene by long wavelength ultraviolet light to an excited state, followed by decay, via hydrogen loss, to a free radical. Radical combination with molecular oxygen results in a hydroperoxide radical which can abstract a hydrogen atom from another hydrocarbon to produce the hydroperoxide and a chain-propagating free radical capable of further reactions. The fluorene hydroperoxide can collapse to fluorene, or cleave heterolitically to an electro-positive hydroxyl, which could then further hydroxylate electron-rich aromatic rings.

It is of interest here to mention another study concerned with chemical oxidation (21), in which the topic of autooxidation is discussed as being differentiated from oxidations due to sunlight. Autooxidation is also a free-radical chain process in the absence of ionic catalysts, with rate of propagation controlled by the rate of hydrogen atom removal by an alkylperoxy radical. Hydrocarbons with tertiary C-H bonds (isoalkanes) will be attacked more readily than other structural types, due to the free-radical stability in the order of tertiary>secondary>primary. This simplistic mechanism is obviously complicated by concomitant losses due to microbial degradation, photooxidation, evaporation, and dissolution.

## Effects of Wavelength, Oil-film Thickness, and Ice/Snow Cover

A very integrative relation exists between the wavelengths of light most effective in inducing oxidation and the absorption of particular wavelengths as a function of film thickness.

One study (82) has been performed to evaluate the relative effectiveness of various wavelengths in the conversion of non-volatile petroleum components to water-soluble compounds. Degradation rates/extents were determined by accumulation (detection) of $CO_2$, sulfuric and acetic acids, organic acids, and esters. Results indicated that longer wavelengths, although possessing lower quantum efficiencies relative to shorter wavelengths (<300 nm), are present in natural sunlight at the highest relative intensities. As depicted in Figure I-12, from Freegarde et al. (82), a slick of 0.1 mm thickness will absorb 90% of all wavelengths less than 600 nm, with absorption rapidly decreasing above 600 nm. It was concluded that since photooxidation reactions are relatively slow (days to weeks to produce significant effects), the actions of sunlight on much thicker slicks

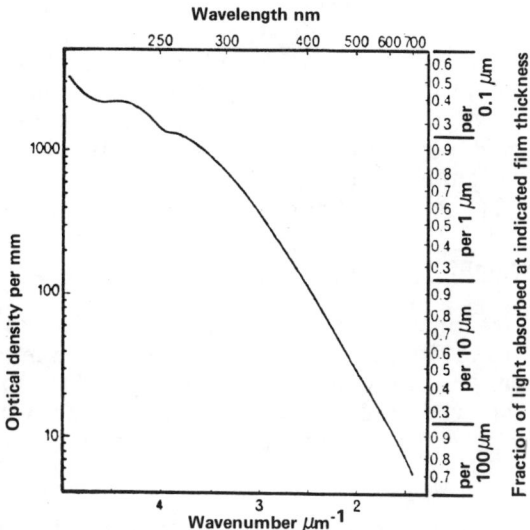

Figure I-12. Optical density vs wavelength and oil film thickness (from Freegarde et al., 82). Reprinted with permission from <u>Laboratory Practice</u>, Vol. 20, No. 1, © by the United Trade Press Ltd.

would be negligible. It was also found that rates of oxidation for slicks less than 0.1 mm in thickness decreased, due to partial transmission of these longer active wavelengths through the film.

Rates of photochemical transformations in relation to film thickness are difficult to assess due to concomitant microbial degradation, dissolution and evaporative losses all occurring in an actual spill situation. One theoretical estimation based on degradation rates for crude oil fractions exposed to artificial light determined a total photochemical degradation period of over three years for a film of 0.4 mm thickness at a rate of 0.07% per day (102). However, rates determined in one study (82) corresponded to decomposition of a 2.5 $\mu$m thickness of oil film per 100 hours (about 2.5 tonnes/km$^2$). With an effective 8 hour day, this corresponds to 0.2 tonne per km$^2$ per day. One other laboratory study suggested facilitation of photo-induced formation of thiocylane-I-oxides from a medium-sulfur crude via thin surface films, which provide for more intimate atmospheric contact (36).

Relatively little is known about the effects of snow and ice cover on the ability of sunlight to induce photochemical degradation. Its relative influence on total petroleum degradation would be expected to be significantly reduced, due to low ambient light intensities (in winter), insulation of oil by the ice/snow cover, and low oil surface areas of exposure (94, 95). Due to the high albedo values for fresh snow (0.85) and wet snow (0.65) in the Arctic region (176), even summer light intensities would most likely result in absorption only as heat energy by oil.

## Inhibition/Enhancement of Photooxidation

Inhibition of photo-induced oxidation occurs via chain-terminating reactions. Organo-sulfur compounds present in the petroleum are oxygenated to sulphoxide products by way of terminating the free radical chain reactions, and thus they inhibit complete oxidation to carboxylic acids (21, 128, 237).

The initial reaction rates may be influenced by the presence of dissolved metal ions of variable valence which act as catalysts. Vanadium, for example, is a common trace metal in petroleum and strongly catalyzes oxidations in the aqueous phase (21, 237).

Photosensitizing compounds, such as xanthone (88), 1-naththol (87, 237) and other naphthalene derivatives (188) have been shown to increase photooxidation rates for petroleum hydrocarbons. Compounds suitable as sensitizers must have strong absorption properties in the visible (or near UV) region, which results in a formation of a singlet or triplet state with a sufficient lifetime and energy to initiate free-radical chain reactions capable of proceeding at low temperatures. Obviously, this compound must also be lipophilic and stable to oxidative processes within the oil-water system.

One study examining the efficiencies of several sensitizing compounds in n-hexadecane photooxidation determined xanthone to be most effective (87). Results suggested the following mechanism (Type I photosensitized oxidation): Light induces formation of triplet state xanthone via intersystem crossing from the excited singlet, which then extracts a hydrogen atom from n-hexadecane (forming a free-radical alkane). The xanthone-hydrogen complex interacts with molecular oxygen to reform the photosensitizer, accompanied by formation of a hydroperoxide radical. This radical then can interact further with other alkanes, which can then combine with molecular oxygen to form peroxides which can decompose to oxygenated radicals. These radicals may then interact with other alkanes to form alcohols. This study also showed that temperature ($2^{\circ}$ vs $25^{\circ}$C) had no significant effect on the rate of alcohol formation. Below is the sequence of events for this process, from Gesser et al. (87):

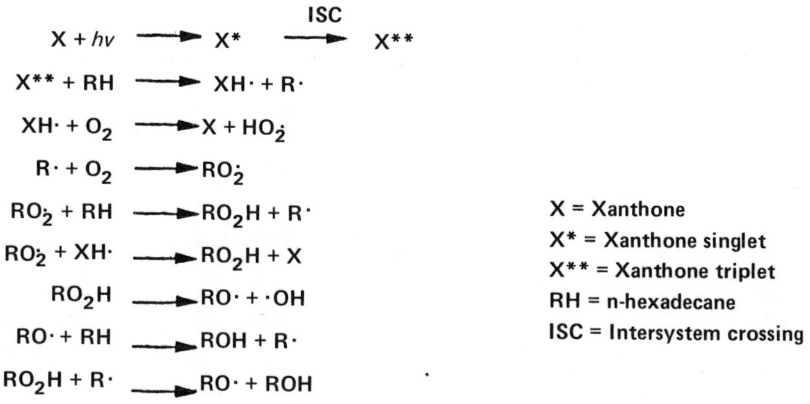

$$X + h\nu \longrightarrow X^* \xrightarrow{ISC} X^{**}$$

$$X^{**} + RH \longrightarrow XH\cdot + R\cdot$$

$$XH\cdot + O_2 \longrightarrow X + HO_2^{\cdot}$$

$$R\cdot + O_2 \longrightarrow RO_2^{\cdot}$$

$$RO_2^{\cdot} + RH \longrightarrow RO_2H + R\cdot$$

$$RO_2^{\cdot} + XH\cdot \longrightarrow RO_2H + X$$

$$RO_2H \longrightarrow RO\cdot + \cdot OH$$

$$RO\cdot + RH \longrightarrow ROH + R\cdot$$

$$RO_2H + R\cdot \longrightarrow RO\cdot + ROH$$

X = Xanthone
X* = Xanthone singlet
X** = Xanthone triplet
RH = n-hexadecane
ISC = Intersystem crossing

Figure I-13. Hypothetical mechanism for sensitizer-induced free-radical oxidation (adapted from Gesser et al., 87). Reprinted with permission from *Environmental Science and Technology*, Vol. 11, No. 6, © by the American Chemical Society.

The effects of photosensitizers on film spreading and photochemical degradation in relation to petroleum composition are discussed below.

Relations to Petroleum Composition
---

Qualitative aspects (i.e. composition) of various petroleums and petroleum products will influence both rates and extents of photo-induced oxidative processes. A study involving exposure of a crude oil surface film to various light sources (102) monitored both film and water column composition with time by IR spectra and GC/GCMS. In the surface film, pristane/n-$C_{17}$ and phytane/n-$C_{18}$ ratios decreased with time. This could be a reflection of greater ease of formation (and stability) of tertiary radicals from the branched alkanes, relative to primary or secondary C-H bonds. More simply, this could also be a reflection of the greater solubilities of isoprenoids over their straight-chain counterparts (see section on dissolution). Identification of the carboxylic acids formed (by GC/MS) from irradiation showed preferential degradation of aromatic compounds. Materials found in the water column showed much greater C=O/$CH_2$ ratios relative to surface film constituents. This indicates that the formation of the acids leads to greater water solubility, as depicted in Figure I-14 from Hansen (102).

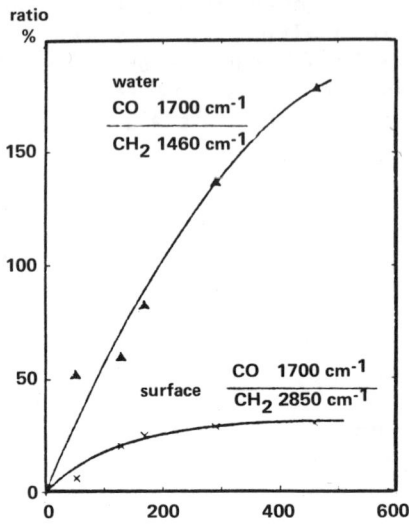

Figure I-14. Increase in oxidation products with time, as given by the IR absorption rations of $C=O/CH_2$ (from Hansen, 102). Reprinted with permission from Marine Chemistry, Vol. 5 (1975), © by the Elsevier Scientific Publishing Co.

Oxygenation or dehydrogenation of alkanes or alkyl-substitutents on cyclic hydrocarbons results in successive formation of alcohols, aldehydes or ketones, and carboxylic acids (which pass into the water column). Conversely, there can be formation of higher molecular weight products via radical polymerization, condensation, or esterification (as discussed previously). Inhibition of diffusion to the oil-water interface due to high oil viscosity and/or low surface area to volume ratios, provides for retention of these higher molecular weight products in persistent tarry residues. Examples of the two types of processes would be degradation of lube oils to products containing acids, alcohols, esters and carboxyl compounds, and gasoline degradation to a "gum" consisting mainly of high molecular weight esters (175).

Dissolution of petroleum components is facilitated by irradiation-accelerated formation of polar organic moieties in the film (12), as well as by surface-active products generated from photooxidation (102). Oil-water emulsions are enhanced by surfactants resulting from chain-breaking reactions in sulphoxide formation and it has been suggested (237) that this process contributes more toward degradation of slicks than the formation of water-soluble oxidation products.

Other products of photooxidation processes include water-soluble metal salts, such as vanadium salts formed by the reaction of vanadyl porphyrins with peroxides (21). Crudes of medium sulfur content (~ 2% by weight) have been shown to be photo-degraded with respect to sulfur compounds, even though sulfur is inhibitory to propagation of the free-radical chain process. One study, utilizing a medium-sulfur crude, allowed prolonged equilibration with seawater and monitored photo-degradation by GC/GCMS (36). Results showed the unresolved complex mixture (UCM) of chromatogram profiles to be almost exclusively a complex mixture of thiocyclane-I-oxides, which are water-soluble oxidation products of indigenous crude oil thiocyclanes. Evidence suggested rapid formation of thiocyclane oxides, given a highly oxygenated surface, atmospheric contact via globular/thin surface layers, and intense illumination. Below is a proposed scheme for hydrocarbon oxidation with respect to thiocyclanes, from Burwood and Speers (36):

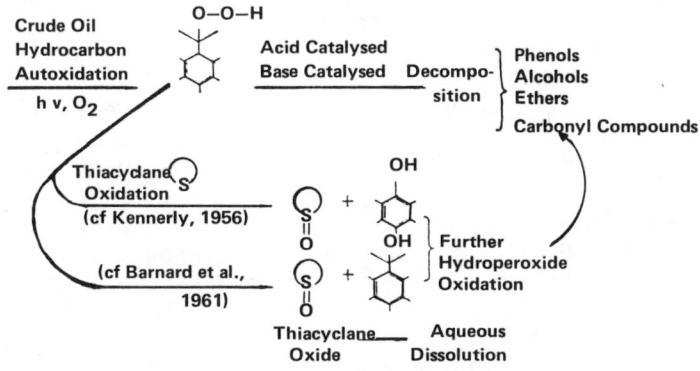

Toxic compounds may be both formed and degraded by photo-induced processes. One study examined phototoxicity effects on primary productivity of a microalga (Phaeodactylum tricornutum) and marine phytoplankton communities by a Kuwait crude oil, as a function of illumination and chemical dispersants (127). Results indicated that photo-induced toxicity was twice that of samples stored in the dark. The dispersant tested (Corexit 8666) enhanced toxicity in both dark and illuminated samples, and provided for resumption of photo-induced toxicity by artificial dispersion of weathered samples which had become inert to photooxidation. Another study demonstrated photochemical formation (short-term UV exposure) of toxic hydroperoxides from benzylic compounds in a Number 2 fuel oil (128, 129). These water-soluble products suppressed growth of yeast cultures at very low concentrations (ca. $10^{-4}$ M) with similar effects by their breakdown products (carbonyl compounds, acids and phenols). However, toxic components indigenous to petroleums may also be degraded via photooxidative processes. Chrysenes and potential carcinogens such as benzanthracenes, benzpyrenes, benzo(a)pyrene and benzo(a)anthracene have been shown to be photooxidized (188).

Photooxidation can also influence the spreading properties of crude petroleums and petroleum products, and these effects must be taken into consideration with regard to the use of photosensitizing agents. One laboratory study has examined spreading properties of n-hexadecane as well as the light and heavy fractions of several crude oils in relation to spreading coefficients (S), lens diameter ($\delta$) and the effects of a sensitizing agent, 1-naphthol (124). Irradiation by artificial sunlight increased spreading coefficients (and consequently lens diameters) for n-hexadecane and the light fraction of one crude, as indicated in Figure I-15, adapted from Klein and Pilpel (124). However, the opposite occurred when the heavy crude fractions were exposed to light, as depicted in Figure I-16, adapted from (124). Addition of 1-naphthol significantly increased spreading for n-hexadecane and the heavy fraction of one crude (which had the lowest viscosity or pour point of all the crudes). The heavy fraction of the other crudes contracted upon addition of 1-naphthol, which is indicated in Figure I-17, adapted from (124). This decrease in lens diameter is believed to be due to the inability of photooxidation products (found at the oil-air interface) to diffuse through the oil and be absorbed at the oil-water interface. Diffusion rates would be inhibited by high viscosities, and may provide for polymerization of oxidation products at the lens surface and subsequent shrinking. Therefore, viscous oils may exhibit different photochemical behavior relative to nonviscous oils and photooxidative effects may actually enhance formation of tarry residues and perhaps tarballs. If photosensitizers are to be added to a spill, addition should be done immediately while the oil is still relatively non-viscous (i.e. before evaporation and dissolution processes can significantly increase viscosities). Conversely, spill cleanup operations such as mechanical surface skimming may actually be facilitated by the contracting effects of photosensitizers on viscous petroleum materials. In this light, the interrelations between oil viscosity, solubility, photoinduced oxidations and use of photosensitizing agents merit further research.

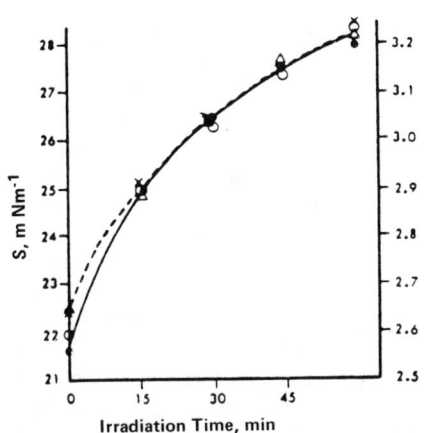

Figure I-15. S and lens diameter δ vs irradiation time for light Tiajuana oil; o = values of S in m Nm$^{-1}$; • = values of δ in cm; x = values of S with $5 \times 10^{-5}$ M 1-naphthol added; and △ = values of δ with $5 \times 10^{-5}$ M 1-naphthol added. Reprinted with permission from Water Research, Vol. 8, © 1974 by Pergamon Press, Ltd.

Figure I-16. Crude oil residues-lens diameter vs irradiation time; O = Tiajuana heavy; • = Peruvian; and x = AMNA Libyan. Reprinted with permission from Water Research, Vol. 8, © 1974 by Pergamon Press Ltd.

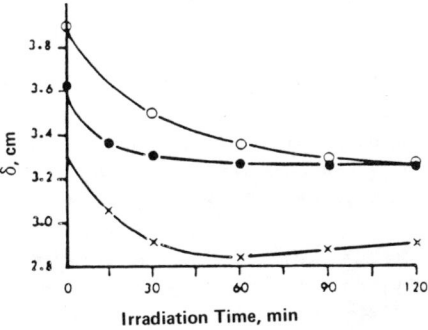

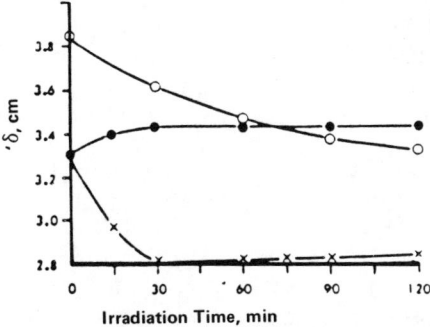

Figure I-17. Crude oil residues containing 1-naphthol: lens diameter vs irradiation time; o = $8.1 \times 10^{-5}$ M 1-naphthol in Tiajuana heavy; • = $5.0 \times 10^{-4}$ M 1-naphthol in Peruvian and x = $1.0 \times 10^{-2}$ M in MNA Libyan. Reprinted with permission from Water Research, Vol. 8, © 1974 by Pergamon Press, Ltd.

## DISPERSION

A dispersion, or oil-in-water emulsion, results from the incorporation of small particles or globules of oil (ranging in size from less than 0.5 micrometers to several millimeters) into the water column. In general, such oil in seawater emulsions are not stable. They can be maintained by continuous agitation, such as that generated by storms or in coastal waters where there may be high turbulence. Milne (162) reported that oil droplets in an oil in water emulsion have an average diameter of 0.5 micrometers, a volume of 6 x $10^{-17}$ liters and a surface area of 8 x $10^{-13}$ meters$^2$. Thus, one milliliter of oil can form up to 16 x $10^{12}$ droplets with a total surface area of about 13 meters$^2$. Due to the inherent instability of these types of emulsions, the oil droplets tend to coalesce and return to the sea surface, forming a slick. Natural or added emulsifiers, detergents, or dispersants, can stabilize such emulsions and the presence of NSO components with hydrophilic side groups such as -COOH (acids), $-CONH_2$ (amids), -OH (alcohols), -CHO (aldehydes), $-OSO_3$ (sulfates) and $-SO_3H$ (sulfonates) in oil can act as such natural surface agents (244). These surfactants have hydrophilic and hydrophobic groups in the same molecule, causing them to orient across the water phase boundary, thus stabilizing the emulsion.

Oil-in-water dispersions can also be stabilized by suspended particulates and Huang and Elliot (110) have examined the effects of naturally occurring suspended solids, such as oxides, hydroxides, carbonates and clays in relation to emulsion stability. Oil-in-water emulsion stability for Nigerian, Venezuelan and Iranian crudes was studied as a function of various concentrations of $Al_2O_3$, $SiO_2$ and Kaolinite. Interactions at the emulsion/solid interface were influenced by the physical and chemical properties of the aqueous phase.

Both the magnitude and sign of the surface charge for the oil droplet and solid particle were important for stabilization. Finely divided solid particles are efficient stabilizers when both the oil droplet and particle have small surface charges of same or opposite sign. However, a complex destabilization phenomenon was observed for particles possessing moderate to large surface charges of opposite sign. This was primarily attributed to particle-particle interactions which tended to exclude the oil.

The results from determinations of electrophoretic mobilities versus pH indicated that at the pH values encountered in marine environments, all three crude oil emulsions (in the absence of particulates), as well as silica, and kaolinite had negative charges, while $\alpha$-$Al_2O_3$ possessed a positive charge. Since most colloids are negatively charged in natural aquatic environments, and the prevailing neutral pH and ionic strength conditions would be inhibitory to formation of relatively large surface charges, it would seem that the mode of stabilization would be the following: when both solid particle and oil droplet have small surface charges of the same sign, the electrostatic forces are weakly repulsive. Therefore, the effectiveness of the solid as a stabilizer will depend on the degree of its hydrophobic character. It was determined that the concentrations of suspended particulates necessary to significantly enhance stabilization should be 100 to 200 mg/l. Such solid concentrations will be primarily found in near-shore environments, or as a result of offshore drilling and/or open ocean dredge disposal.

Generally, oil begins to be dispersed immediately after it is discharged and the dispersion process is most significant during the first ten hours (237). Once formed, the particles can then continue to break down and/or disperse throughout the lifetime of the spill. By one hundred hours, dispersion has usually overtaken spreading as the primary mechanism for distributing the spilled oil about its center of mass.

Following the spill of Bunker C fuel oil from the Arrow in Chedabucto Bay, Nova Scotia, Forester (79) assessed the distribution of oil particles in the water column. The estimated length ($\sqrt[3]{vol.}$) of the droplets to a depth of 80 meters ranged from five micrometers to one millimeter, however, occasionally droplets as large as two millimeters were found. The droplets were formed primarily by the action of surf and waves, and under open water conditions, such turbulence generally plays a major role in controlling the dispersion of particle formation and suspension. Forester (79) estimated that the breakup of oil particles could be related to the turbulent energy spectrum according to the following equation: $L^2 x P(L)/E(\lambda)$ = a constant (independent of L). Where L is particle length; P(L) is the probability per unit time that a particle size L will bisect; and $E(\lambda)$ is the spectral density of turbulent energy of wave length required to break the particle. The $L^2$ term indicates that this model is adequate only for spherical particles and the equation suggests that similar, possibly more abundant, dispersed oil particles would be associated with lower turbulent energy areas. Analysis of the samples taken at the time of the spill indicated that the total oil concentration decreased with depth while the relative abundance of smaller particles increased. It should be noted that when seawater is the continuous phase of emulsified oil in seawater, there is really no limit to the dispersion of the oil droplets. A large population of small oil particles (less than one millimeter in diameter) was formed during the first 15 days following the Arrow spill. This was attributed to 1) stormy weather conditions; 2) the presence of about ten tons of the dispersant "Corexit 8666", which was sprayed from an aircraft onto the bay during the first few days following the spill in an unsuccessful attempt to emulsify the oil; and 3) the unweathered condition of the oil (79). Following the spill, the oil particles could be traced from the Arrow approximately 250 kilometers southward from Chedabucto Bay in a band extending up to 25 kilometers offshore. Two weeks after the wreck, particles were still found 70 kilometers to the east of Nova Scotia in a 10-kilometer-wide tongue, probably carried there by wind driven currents.

In a related laboratory study, Prouse (97) examined the formation of dissolved and particulate phases of three different oils (Venezuelan crude oil, Number 2 fuel oil and Bunker C fuel oil) in natural seawater. The oils were dissolved in hexane and then mixed with seawater with vigorous shaking. The size distribution of the suspended or dispersed droplets was then determined with a Coulter counter. In addition, fluorescent spectrometry was used to determine the total oil and subparticulate oil in the solutions. The total oil content, which included the dissolved phase and the suspended particulates in the seawater for the three oils, was about the same (0.3 mg/l), but the ratio of the particulate to dissolved or subparticulate oil varied considerably. This ratio of particulate to subparticulate fractions was related directly to the viscosity of the oils: the Bunker C fuel oil showed a particulate/subparticulate ratio of 40.7; the Number 2 fuel oil a ratio of 9.8; and the Venezuelan crude oil, a ratio of 7.5 (97). Gordon and his coworkers also observed that only 2-13% of the oil in the seawater passed through an 0.8 micron diameter filter, and that approximately two-thirds of the oil which was retained on the filter had particle sizes greater than 2 micrometers in diameter.

In preparing their Number 2 fuel oil-saturated water samples (oil-in-water dispersion) for the controlled marine ecosystem program at the University of Rhode Island, Gearing et al. (85) noted that the composition and physical form of the oil in water dispersion depended to a great extent on the amount of energy applied in creating the mixture. When the mixing was energetic, the aqueous mixture had a greater resemblance to the original oil and there was little or no enrichment of the more soluble hydrocarbons which comprised 30-40% of the total lipids in dissolved fraction used in the MERL experiments, versus 20-25% of the lipids in the original oil. No such fractionation of the aliphatic fractions was observed. In the MERL experiments the oil was suspended as micelles, or microdroplets ranging from less than 0.3 to 6 microns in diameter, with the greatest number of particulates being less than 0.3 microns. These data agree with those of Forrester, showing that an oil-water dispersion is not a true solution, but rather, an accommodated form of small droplets and soluble fractions. Filtration of the water through a 45-micron sieve and a 0.3 micron glass fiber filter demonstrated that 82% of the oil was associated with the fraction having droplet sizes greater than 0.3 microns, but less than 45 microns.

Unless chemical dispersants are added or the oil contains large concentrations of surface active NSO components or high levels of dibenzothiophene, which may be photochemically oxidized to sulphoxides, most oil in water dispersions will tend to coalesce back to a water-in-oil emulsion where further weathering is then retarded. In the oil-in-water state, the surface area of the oil is greater, and it is, therefore, subject to greater dissolution. Eventually, however, this process becomes self-limiting and only the more viscous residues will remain. These then tend to agglomerate and revert to the water-in-oil emulsion or mousse.

DISSOLUTION

Rates of dissolution for the various components of a petroleum slick depend on rather complex interactions between properties inherent to the oil (i.e. molecular structure of compounds and relative abundance of these components) and the physical-chemical properties of the immediate environment (i.e. salinity, temperature, etc.). Not only does this complex interaction of compositional and environmental factors exist for rates of evaporation, the overall rate of slick disappearance depends on interactions between evaporation and dissolution processes. One study has examined rates of disappearance for aliphatic and aromatic components from small slicks on seawater (103).

Measurements of slick components indicated the dissolution rate to be less than half of the evaporation rate. A mathematical model designed in the study indicated that rates of dissolution may be as low as 1% of evaporation rates.

Many studies have provided data to define solubility as a function of molecular structure. Principal determinants of solubility for any particular petroleum hydrocarbon include the molecular volume (expressed as $cm^3$/mole) and the presence of "active" groups (e.g. aromatic rings or olefinic bonds). Solubility is generally inversely proportional to molar volume (74, 81, 141, 150, 152), which is a linear function of carbon number. Roughly, the solubility decreases by a factor of three per carbon number (82), but linearity of this relation falls off for n-alkanes above n-$C_{10}$. These longer chain paraffins exhibit greater solubility than would be expected by extrapolation (81, 152), with the decrease in solubility for each additional $CH_2$ being considerably smaller than for shorter chain lengths. However, the true solubilities of n-paraffins greater than n-$C_{10}$ may be decreased due to formation of hydrocarbon aggregates of micelles arising from impurities in the water phase. Results of one study suggested that n-$C_{11}$ may represent a change from a state of true solubility to one of accommodation (aggregation via intermolecular associations) between n-$C_9$ and n-$C_{10}$ (152).

Branched alkanes demonstrate greater solubilities for a given carbon number than their straight-chain counterparts (for paraffins, olefins, and acetylene hydrocarbons), and this seems to be due to increased vapor pressures relative to corresponding compounds, as opposed to a structural function (82, 150, 151). Exceptions to this trend are methane and neopentane, which have lower solubilities than predicted (151).

Ring formation also enhances solubility for a given carbon number or molar volume (150, 151). For example, cyclohexane and toluene have similar molar volumes (108.1 $cm^3$/mole and 106.3 $cm^3$/mole, respectively) but drastically different solubilities (55 $g/m^3$ and 515 $g/m^3$, respectively) in aqueous systems (141).

The degree of saturation is inversely proportional to solubility for both chain and ring structures. The addition of a second or third double bond increases solubility proportionately, and it has been shown that presence of a triple bond increases solubility to a greater proportion than presence of two double bonds (151). Therefore, the most water-soluble petroleum hydrocarbons will be those with the lowest molar volume and greatest aromatic/olefinic character.

Solubilities can also be affected by the type and number of substituents present in the molecular structure. Studies on solubilities of polychlorinated biphenyls (PCB's), although not found in crude petroleums, demonstrate an inverse relation between solubility and degree of chlorination (63, 64). Increased adsorption onto particulate materials may be a feasible explanation for decreased solubility with increased chlorination, as is evidenced by results from a model study (64). This model represents an attempt to define accumulation of stable organic compounds on marine particulate interfaces by way of an equilibrium sorption mechanism. Although the model needs further refinement in relation to selected abiotic environmental parameters, it has promise in defining relative concentrations of non-polar organic compounds in seawater and suspended particulates.

Another study examining solubilities of alkylbenzenes in distilled water and seawater (216) attempted to determine the effects of alkyl-substituent size and position on solubilities for mono- and polysubstituted structures. Figure I-18 from Sutton and Calder (216) demonstrates the linear relation between the natural log of solubility and molar volume for the alkylbenzenes studied. However, in accordance with Henry's law, some of this linear relation may be due to differences in equilibrium vapor pressures between aromatics (smaller aromatics have higher vapor pressures than larger species). The solubility data from distilled water was normalized to one atmosphere equilibrium vapor pressure, as indicated in Figure I-19 from (216). The differences between the two figures indicate that the size of the alkyl-substituent is important in determining solubility of monosubstituted structures. For polysubstituted alkylbenzenes, the positioning of substituents is also important.

The presence of co-solutes (i.e. hydrocarbon mixtures) can influence the solubility of a particular hydrocarbon, relative to the binary system of pure compound and liquid phase. However, no general trends are apparent, as solubilities may increase or decrease for various hydrocarbons with additional system components (74).

An inverse relation exists between salinity and hydrocarbon solubilities for both aliphatic and aromatic components (74, 82, 214, 216), with an approximate decrease of 30% for n-paraffins between fresh and seawater (82). Table I-5

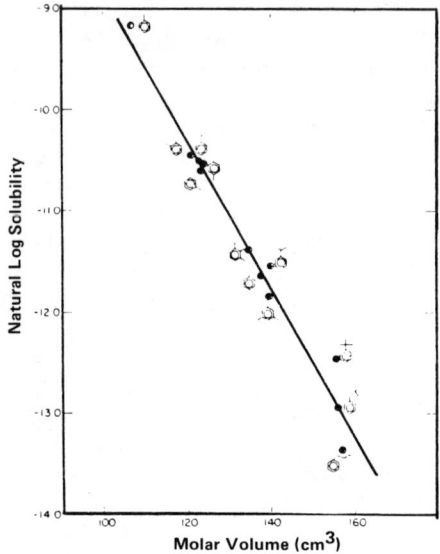

Figure I-18. Natural log of solubility (as mole fraction) vs molar volume (from Sutton and Calder, 216). Reprinted with permission from the Journal of Chemical and Engineering Data, Vol. 20, No. 3, © by the American Chemical Society.

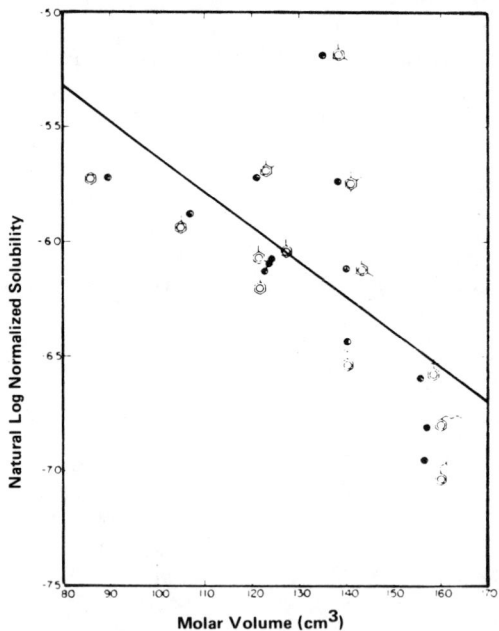

Figure I-19. Natural log of normalized solubility (as mole fraction) vs molar volume (from Sutton and Calder, 216). Reprinted with permission from the Journal of Chemical and Engineering Data, Vol. 20, No. 3, © by the American Chemical Society.

Table 1-5. Solubilities of aliphatic and aromatic petroleum hydrocarbons in seawater and distilled water at 25°C (adapted from Sutton and Calder, 214,216). Reprinted with permission from the Journal of Chemical and Engineering Data, Vol. 20, No. 3, © by the American Chemical Society.

| Compound | Solubility in Distilled Water | Solubility in Seawater |
|---|---|---|
| Dodecane ($C_{12}$) | 3.7 (ppb) | 2.9 (ppb) |
| Tetradecane ($C_{14}$) | 2.2 | 1.7 |
| Hexadecane ($C_{16}$) | 0.9 | 0.4 |
| Octadecane ($C_{18}$) | 2.1 | 0.8 |
| Eicosane ($C_{20}$) | 1.9 | 0.8 |
| Hexacosane ($C_{26}$) | 1.7 | 0.1 |
| Toluene | $534.8 \pm 4.9$ (ppm) | $379.3 \pm 2.8$ (ppm) |
| Ethylbenzene | $161.2 \pm 0.9$ | $111.0 \pm 1.3$ |
| $o$-Xylene | $170.5 \pm 2.5$ | $129.6 \pm 1.8$ |
| $m$-Xylene | $146.0 \pm 1.6$ | $106.0 \pm 0.6$ |
| $p$-Xylene | $156.0 \pm 1.6$ | $110.9 \pm 0.9$ |
| Isopropylbenzene | $65.3 \pm 0.8$ | $42.5 \pm 0.2$ |
| 1,2,4-Trimethylbenzene | $59.0 \pm 0.8$ | $39.6 \pm 0.5$ |
| 1,2,3-Trimethylbenzene | $75.2 \pm 0.6$ | $48.6 \pm 0.5$ |
| 1,3,5-Trimethylbenzene | $48.2 \pm 0.3$ | $31.3 \pm 0.2$ |
| $n$-Butylbenzene | $11.8 \pm 0.1$ | $7.09 \pm 0.07$ |
| $s$-Butylbenzene | $17.6 \pm 0.2$ | $11.9 \pm 0.2$ |
| $t$-Butylbenzene | $29.5 \pm 0.3$ | $21.2 \pm 0.3$ |

lists the solubilities for some aliphatic (214) and aromatic (216) petroleum hydrocarbons in distilled water and seawater (35°/oo + 0.5) at 25°C. For the paraffins, the magnitude of this "salting-out" effect is directly proportional to the molar volumes in accordance with the McDevit-Long theory (214), which attributes salting in or salting out to the effect of electrolytes upon water structure.

Table I-6 lists solubilities for aliphatic and aromatic petroleum hydrocarbons in distilled water and filtered seawater (35°/oo). In addition to providing additional examples of the negtive effect of salinity on compound solubility, it can be seen that, even in seawater, the carcinogenic/toxic polynuclear aromatics (e.g. phenanthrene) are much more soluble than the nontoxic, long chain n-paraffins.

Dissolved organic matter (DOM) in the marine environment enhances solubility, due to its surface-active nature. One study, utilizing natural marine water and NaCl solutions, examined the effect on various hydrocarbon solubilities due to removal of the dissolved organic matter (28). A 50 to 99% decrease in the amounts solubilized was observed for n-alkanes and isoprenoids, with the decreases being directly proportional to the amount of DOM removed (by activated charcoal and UV oxidation). However, the aromatics examined (anthracene, phenanthrene, and dibutylphthalate) were unaffected by this process. These surface-active materials (DOM) were also shown to reinstate the expected solubilities in organic depleted seawater. This study also provided evidence which suggested the mode of solubilization via incorporation into micelles formed by intermolecular association of the surface-active humic type monomers. These marine humic substances may comprise up to 90% of the DOM in seawater, and provide for hydrocarbon solubilization into a semi-colloidal or micellar state. The influence of pH on solubilities was examined, as the presence of ionic species is necessary for micelle formation. Solubilities for n-alkanes increased with pH in a fulvic acid (isolated from marine sediment) and saline (NaCl) solution, with maximum solubilities observed for ionic strength of 0.3.

Another study has examined the association of hydrocarbons and marine particulates in saline solutions with respect to dissolved organic materials and temperature (162). The organic materials (humic substances) associated with the various clays and sediments tested provided for decreased particulate association, and hence increased solubilities for two alkanes (hexadecane and eicosane) and two aromatics (anthracene and phenanthrene). Temperature effects were

Table I-6. Additional solubility data on a variety of petroleum hydrocarbons in distilled water and filtered seawater (from Clark and MacLeod, 48). Reprinted with permission of Academic Press, Inc., © 1977.

| Compound | Carbon Number | Solubility[a] (ppm) | Reference |
|---|---|---|---|
| **PARAFFINS** | | | |
| Methane | 1 | 24 | 152 |
| Ethane | 2 | 60 | 152 |
| Propane | 3 | 52 | 152 |
| n-Butane | 4 | 61 | 152 |
| n-Pentane | 5 | 39 | 152 |
| n-Hexane | 6 | 9.5 | 152 |
| 2-Methylpentane | 6 | 13.8 | 152 |
| 3-Methylpentane | 6 | 12.8 | 152 |
| 2,2-Dimethylbutane | 6 | 18.4 | 152 |
| n-Heptane | 7 | 2.9 | 152 |
| n-Octane | 8 | 0.66 | 152 |
| n-Nonane | 9 | 0.220 | 215 |
| n-Decane | 10 | 0.052 | 215 |
| n-Undecane | 11 | 0.0041 | 215 |
| n-Dodecane | 12 | 0.0037 | 215 |
| | | 0.0029 (SW) | |
| n-Tetradecane | 14 | 0.0022 | 215 |
| | | 0.0017 (SW) | |
| n-Hexadecane | 16 | 0.0009 | 215 |
| | | 0.0004 (SW) | |
| n-Octadecane | 18 | 0.0021 | 215 |
| | | 0.0008 (SW) | |
| n-Eicosane | 20 | 0.0019 | 215 |
| | | 0.0008 (SW) | |
| n-Hexacosane | 26 | 0.0017 | 215 |
| | | 0.0001 (SW) | |
| n-Triacontane | 30 | 0.002[b] | 172 |
| n-Heptacontane | 37 | 0.001[b] | 185 |
| **CYCLOPARAFFINS** | | | |
| Cyclopentane | 5 | 156 | 152 |
| Cyclohexane | 6 | 55 | 152 |
| Cycloheptane | 7 | 30 | 152 |
| Cyclooctane | 8 | 7.9 | 152 |
| **AROMATICS** | | | |
| Benzene | 6 | 1780 | 152 |
| Toluene | 7 | 515 | 152 |
| o-Xylene | 8 | 175 | 152 |
| Ethylbenzene | 8 | 152 | 152 |
| 1,2,4-Trimethylbenzene | 9 | 57 | 152 |
| iso-Propylbenzene | 9 | 50 | 152 |
| Naphthalene | 10 | 31.3 | 73 |
| | | 22.0 (SW) | |
| 1-Methylnaphthalene | 11 | 25.8 | 73 |
| 2-Methylnaphthalene | 11 | 24.6 | 73 |
| 2-Ethylnaphthalene | 12 | 8.00 | 73 |
| 1,5-Dimethylnaphthalene | 12 | 2.74 | 73 |
| 2,3-Dimethylnaphthalene | 12 | 1.99 | 73 |
| 2,6-Dimehtylnaphthalene | 12 | 1.30 | 73 |
| Biphenyl | 12 | 7.45 | 73 |
| | | 4.76 (SW) | 73 |
| Acenaphthene | 13 | 3.47 | 73 |
| Phenanthrene | 14 | 1.07 | 73 |
| | | 0.71 (SW) | |
| Anthracene | 14 | 0.075 | 172 |
| Chrysene | 18 | 0.002 | 172 |

[a] In distilled water, except where noted by (SW), indicating filtered seawater, usually corrected to a salinity of 35 °/oo (parts per thousand); ppm = μg/g.
[b] Extrapolated.

also examined, and data suggested that as hydrocarbon solubility increases with temperature, less association with clays or particulates occurs. The heats of sorption onto Bentonite clay calculated for eicosane and anthracene indicated physical adsorption of the van der Waals type (<10 kcal/mole).

The relations of petroleum composition to overall solubility in seawater involve not only properties inherent to the oil itself, but environmental conditions that will influence partitioning of petroleum components, in terms of relative contributions by evaporation, photochemical processes, microbial degradation, sedimentation, etc. At this time it is not possible to derive anything but very general comments on the relationship between petroleum composition and dissolution processes. Highly viscous oils will most likely be more resistant to this process of slick degradation: more viscous oils may result in slicks with greater film thickness. Thick films will retard dissolution due to increased diffusion distances for components to reach the oil/water interface. Relatively light petroleums will experience decrease in solubility as weathering processes provide for the loss of more volatile (and more soluble) components. As is evident from the sections on photochemical and microbial degradation, properties inherent to petroleums influence these processes which in turn can influence solubilities.

EMULSIFICATION

The process of emulsification represents a change of state from an oil on water slick or an oil-in-water dispersion to a water-in-oil emulsion with eventual possible formation of a thick, sticky mixture which may contain up to 80% water (21, 22). A greater than 80% water-in-oil emulsion ("mousse") has a viscous chocolate-like constituency, while 50-80% water-in-oil emulsions have a grease-like consistency. Emulsions with 30-50% water-in-oil are fluid and resemble the parent oil (154). This process may proceed fairly rapidly and may occur as a result of natural forces (110). The turbulent energy provided by a gently rolling sea-state is sufficient for water in oil emulsification. Ease of formation and stability are primarily related to the chemical composition of the petroleum.

Environmental parameters such as water temperature, salinity, pH, presence of surface-active agents, and suspended particulate load seem to have predominant effects on oil-in-water emulsions (dispersions), as opposed to water-in-oil emulsions (21, 110). One study has indicated the formation of a mousse to be independent of water salinity, with little significant influence by bacterial action or suspended debris (22). Another study, however, has suggested possible stabilization by external agents such as bacterial slime, plankton, or organic debris (21).

Several physical parameters inherent to an water-in-oil emulsion have been examined in relation to stability (22). A correlation has been found between the size of encapsulated water droplets (generally <10 μm diameter), with emulsions containing relatively large droplets being the least stable. In that work, other parameters such as specific gravity, kinematic viscosity, and pour point did not demonstrate significant relationships to stability.

This study (22) also examined chemical parameters in relation to ease of formation and stability for water-in-oil (mousse) emulsions. Factors not demonstrating significant correlations included acidity, sulfur content, and wax content. These results seem to be in conflict with both the effects of sulfur on photo-induced production of surface-active materials (note section on photooxidation), and correlations between high wax content and mousse and tarball composition. Those factors showing definite (positive) correlations were the percent residue boiling over $700^\circ F$, asphaltene content, and vanadium content. Crude petroleums with a high asphaltene content (e.g. Venezuelan and Kuwait crudes) will therefore tend to form mousse emulsions with greater ease and stability than highly paraffinic oils, such as crudes from Libya and/or Nigeria (67). Stability is related to complex chemical components in non-volatile residues, particularly in asphaltenes, and possibly may be related to the presence of porphyrins (including vanadium complexes). Surface-active materials seem to be involved in formation of the water-in-oil mousse, with asphaltenes and metallo-porphyrins being the most probable sources. Once emulsions become stranded on shore lines, they will interact with sand, debris, and organic detritus to form tarry masses. Stranding in the high littoral zone will allow evaporation of the encapsulated water, which results in significant decreases in degradation rates and the development of crusty outer layers.

TARBALL FORMATION

The formation of tarry residues, or tarballs, is facilitated by way of the cumulative effects of evaporation, dissolution, microbial degradation, and photooxidative processes. The first three types of processes interact along with environmental conditions to result in formation of a stable mousse (w/o) emulsion. At this point, microbial effects probably become insignificant due to limited diffusion of oxygen and/or minerals and nutrients into the mousse (110), and overall weathering rates decrease significantly. The results of one study, involving a combination of an evaporative weathering model and analyses of pelagic tar lumps, indicated the principal processes affecting petroleum residues on the ocean surface to be evaporation and physical fragmentation of the mousse (57). Subsequent alterations in the composition of the mousse may also be due to photooxidative processes by way of polymerization and/or condensation reactions (note previous section on photooxidation). However, photochemical changes will most likely cease to be important in the tarry residues or tarballs, as the opaque nature of the tar inhibits transmission of active wavelengths.

## AGGLOMERATION, SEDIMENTATION AND SINKING OF PETROLEUM AFTER RELEASE INTO THE MARINE ENVIRONMENT

As in the other weathering induced degradation processes, the agglomeration and sinking of petroleum hydrocarbons are caused by a variety of phenomena working in concert. Generally, increases in density lead to sinking, and these changes are usually due to 1) evaporation and dissolution of the lower molecular weight compounds, 2) photooxidation and subsequent dissolution of oxygenated materials, 3) the formation of viscous and higher density water in oil emulsions, 4) the incorporation of particulates and the agglomeration of oil particulate mixtures, and 5) the incorporation of higher density microorganisms and even macroorganisms such as barnacles.

The capacity for certain oils to sink as a function of their weathering is also a function of the parent petroleum composition itself. For example, Bunker C or Number 6 fuel oil, which has a density greater than most other products can sometimes sink without extensive degradation (50). After two tankers collided in 1971 at the center of the Golden Gate in the entrance of San Francisco Bay, a major spill of Bunker C fuel oil occurred. Some of the oil globules were neutrally buoyant and were rapidly incorporated into the near-bottom water of the Bay. This was demonstrated by the oil-laden bottom waters moving independently of the surface floating oil slicks, resulting the deposition of oil on landward beaches, which were 16 kilometers east (upriver) of the eastern-most limit of beaches stained by the floating oil.

Generally, however, extensive modification of oil occurs before it is sedimented. For example, the residue (greater than $520^{\circ}C$ bp) fraction of Kuwait crude has a specific gravity of 1.023, compared to 0.869, the specific gravity of the starting material. In that seawater has a specific gravity of 1.025, heavily weathered residues can disperse by sinking processes.

In test tank studies using Number 2 fuel oil at the Marine Ecosystem Research Laboratory at the University of Rhode Island, fractionation of the lower molecular weight aromatic compounds (including up to 3-ring aromatics) into the dissolved phase occurred before partitioning of the oil onto suspended particulate matter and subsequent sinking (85).

Several authors (85, 113, 162, 237, 238, 248) have indicated that particulate oil interactions are a dominant process in the ultimate deposition of petroleum. The density of seawater is 1.025 grams/cm$^3$, so only limited amounts of detrital mineral or particulate organic material (specific gravity 2.5 - 3.0 are common) are required to sediment several times their volume in oil. The nature of this sedimentation is, of course, a function of the oil composition as well as the type of particulates (61, 85, 143, 162, 217, 238, 248).

As noted above, some weathering and fractionation of the oil before incorporation onto suspended particulate material appears to be common. This was found in test tank studies (85) where the aromatic/aliphatic ratio in the sediment was much lower than that in the parent oil, suggesting some differences in the mechanism and/or rate of deposition throughout the ecosystem. Specifically, two to thirty-four percent of the high molecular aliphatic acyclic and greater than 3-ring hydrocarbons were adsorbed to the suspended particulates and sediments in contrast to 0.1% of the naphthalene and methylnaphthalene compounds which were the predominant aromatic materials in the Number 2 fuel oil used in the test.

Winters (238) reported similar partitioning in a study of the fate of petroleum derived aromatic compounds in seawater held in outdoor tanks as part of the South Texas Outercontinental Shelf Study sponsored by the Bureau of Land Management. Using gas chromatography and GC/MS on two simulated oil spills and one mixture of aromatic compounds added to a test tank, Winters showed that the concentration of petroleum derived alkanes was approximately ten times greater in the particulate fraction. Aromatic compounds were at least five times more concentrated in the dissolved fractions. The individual aromatic components themselves also showed differential partitioning with the parent aromatic and monomethyl compounds being enriched in the dissolved fraction and the more highly alkylated aromatics being enriched in the particulate fractions. Again, in this experiment most of the more volatile components including many of the naphthalenes were lost and were not found in the water column or on the particulates under the experimental conditions employed.

Similar partitioning of petroleum hydrocarbons onto particulates has been observed in the near coastal regions and open ocean in non-spill situations (61, 185). In samples of particulate matter collected along transects perpendicular to the south Texas outercontinental shelf near Corpus Christi, Parker et al. (185) noted that total particulate hydrocarbon burdens decreased with increasing distance from shore. The concentration of the higher molecular weight ($n-C_{28}$ through $n-C_{30}$) compounds, however, remained relatively constant. These authors concluded that this distribution could be explained by introduction of the hydrocarbons near shore with subsequent movement of particulate bound hydrocarbons offshore, with preferential retention of the higher molecular weight hydrocarbons during weathering. These authors also detected higher molecular weight polycyclic aromatic compounds on the particulate material. Concentrations of these materials were too low for quantitation; however, they could be detected by selected ion monitoring GC/MS. Among the polynuclear aromatic compounds identified were alkyl-naphthalenes, phenanthrenes, dibenzothiophenes, fluoranthene, and pyrene.

In a study designed to measure the partitioning of petroleum hydrocarbons among seawater, particulates and the filter feeding mussel, Mytilus californianus, deLappe, et al. (61) found a distinct partitioning of the higher molecular weight petroleum components ($>n-C_{22}$ to $n-C_{35}$) on the particulates in the near coastal zone, while the lower molecular weight components ($n-C_{12}$ to $n-C_{22}$) were present in the dissolved phase. Figures I-20 and I-21 present chromatograms showing the relative concentrations of aliphatic and aromatic hydrocarbons in the dissolved and particulate matter samples collected near a natural oil seep at Goleta Point, California 25-26, August 1978 and off the coast of Corona del Mar, California, 14-15 September 1978, respectively (61). Similar partitioning of lower and higher molecular weight chlorinated hydrocarbons between the dissolved and particulate phase has been reported by Risebrough (194), and Dexter and Pavlou (64).

The nature of the particulate material is also critical in controlling the extent of oil particulate adsorption and sinking (143, 162, 217, 248).

ABIOTIC FACTORS/PROCESSES 47

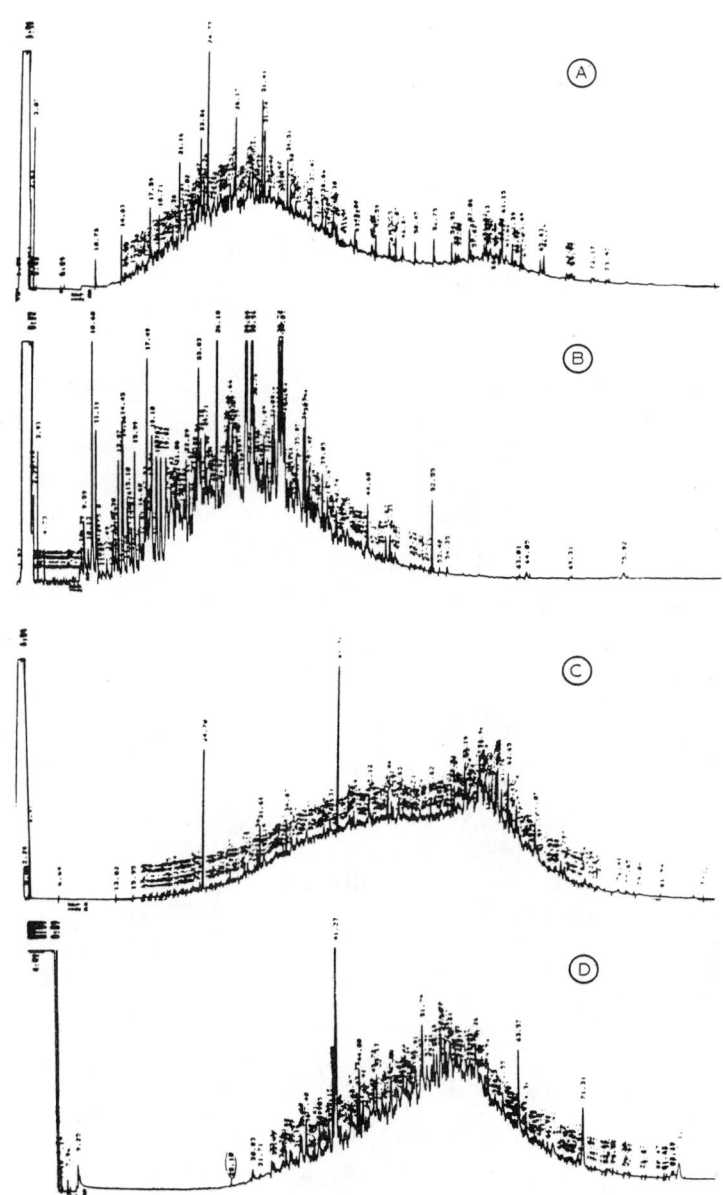

Figure I-20. Glass capillary FID gas chromatograms of seawater extracts (A = alophatic fraction and B = aromatic fraction) and suspended particualte material extracts (c = aliphatic fraction and D = aromatic fraction) collected simultaneously off Goleta Point, CA, August 1978 (from deLappe et al., 61).

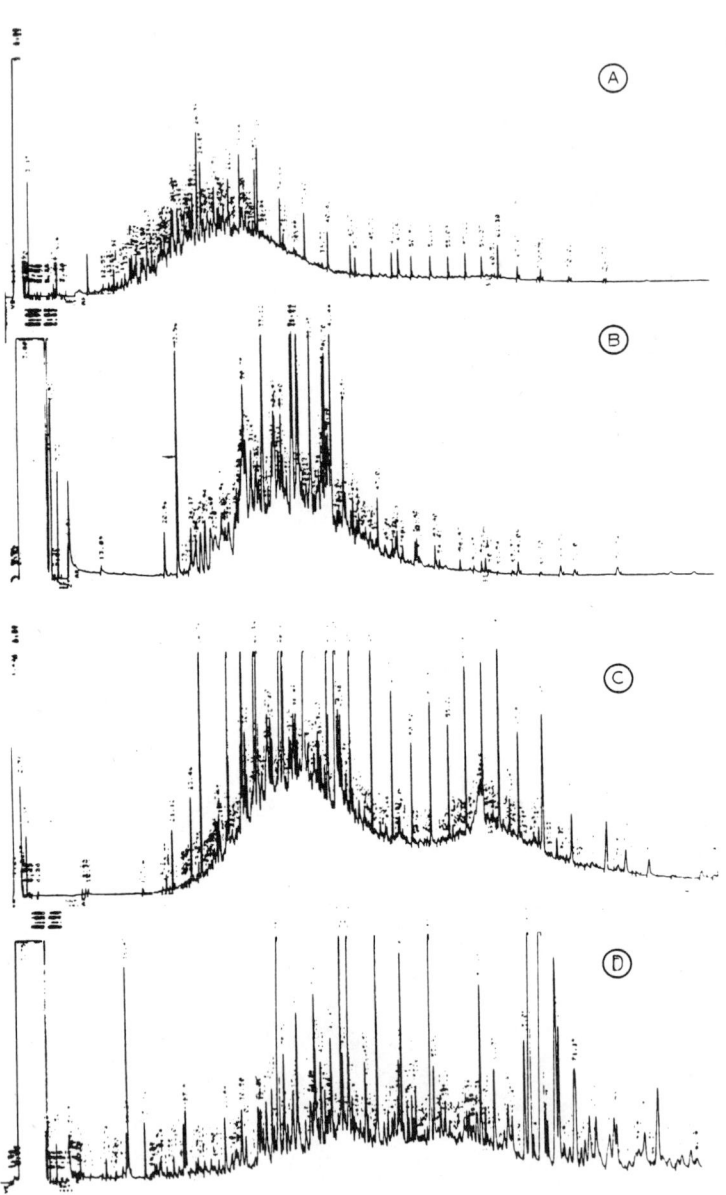

Figure I-21. Glass capilalry FID gas chromatograms of seawater extracts (A = aliphatic fraction and B = aromatic fraction) and suspended particulate matter extracts (C = aliphatic fraction and D = aromatic fraction) collected simultaneously of Corona del Mar, CA, September 1978 (from deLappe et al., 61).

In laboratory studies examining the efficiency of different particles for oil adsorption, Meyers and Quinn (162) found that the adsorption efficiency for the less than 44 micron particle size fraction, decreased in the order of kentonite>kaolinite>illite>montmorillonite. In tests with Number 2 fuel oil and these minerals it was found that in general the resolved components were preferentially adsorbed, compared to the more polar materials in the unresolved complex mixtures. The nature of the physical interaction or adsorption was believed to be primarily due to van der Waals type attractions (less than 10 kcals/mole) as determined by heat of adsorption experiments for eicosane and anthracene with bentonite clay.

To determine the extent of organic material on the sediment and its influence on adsorption, sediment samples from Narraganset Bay were treated with 30% peroxide to remove any indigenous organic matter (162). The less than 44 micron sized fraction of this treated sediment was then exposed to oil samples. The non-hydrogen peroxide treated control sediments sorbed about one-third less Number 2 fuel oil components than the same sediments that were pre-treated with the hydrogen peroxide to remove organic coatings. Meyers and Quinn (162) stated that the organic matter (which was presumed to be humic substances) presumably masked the sorption sites in the sediment, thereby reducing the available surface area for sorption of the oil components. Suess (217) on the other hand, has suggested that a 3-4% organic material coating on clay will enhance the sorption process by providing, in effect, a lipophilic layer to provide a non-polar hydrophobic binding site. Meyers and Quinn (162) also found that the sorption was apparently increased with increases in salinity and that it decreased with temperature. Also, their results indicated that the sorption process was not completely reversible in that only 15% of the fuel oil adsorbed onto clay particles was removed after three rinses.

In a more recent study, Zürcher et al. (248) completed laboratory studies to consider the dissolution, suspension, agglomeration and adsorption of fuel oil onto particulates in a specific laboratory study which precluded evaporation. They used kaolinite as the particulate material and a fuel oil which contained primarily alkanes and aromatics in the benzene to $n-C_{24}$ range, with the maximum n-alkanes occurring around $n-C_{13}$. Their quantifications were done by IR spectroscopy and gas chromatography. Examination of the fraction dissolved in water showed almost exclusively low molecular weight aromatics in the benzene to methylnaphthalene range, with Kovat indices 600 to 1300 (a Kovat index of

600 corresponds to $n-C_6$; 700 - $n-C_7$; etc.). The adsorbed fraction contained n-alkanes from $n-C_{14}$ through $n-C_{32}$ with the maximum occurring between $n-C_{14}$ and $n-C_{21}$. Figure I-22 shows the chromatograms and IR spectra of the original oil, the water soluble fraction showing primarily the lower molecular weight aromatics, the adsorbed fraction showing the higher molecular weight component and agglomerate fraction, which was produced by vigorous mixing causing the oil globules to physically interact with the kaolinite clays. Their data clearly support the field work of deLappe et al. (61) and Parker (185), which showed the partitioning of the lower molecular weight hydrocarbons into the water column and the higher molecular weight hydrocarbons onto the suspended particulate material in natural coastal environments.

As the chromatograms illustrate, the particulates clearly adsorb to the higher molecular weight hydrocarbons, more efficiently than the lower molecular weight components. This agrees with the Traube rule that homologous series of organic substances adsorb more rapidly from water with increasing molecular weight. The clay minerals in this experiment contained about 200 milligrams of hydrocarbon per kilogram of dry material after ten hours. (Meyers and Quinn [162] reported a similar value of 162 mg/kg for dry kaolinite.) This concentration did not change in up to 100 hours of additional exposure time. When the agitation of the oil-kaolinite-water mixture increased and oil droplets became dispersed in the mixture, the oil-kaolinite agglomerates settled rapidly to the bottom of the flask. The chromatogram of this oil clearly shows that it is similar to original starting material, except that some of the lower molecular weight materials have been lost, presumably from volatilization during sample processing. The oil load on kaolinite during more vigorous agitation was 20 grams of oil per kilogram of dry material after 20 hours exposure. This value was 100 times greater than the amount adsorbed out of solution. With the rapid sedimentation of the oil-kaolinite mixture, the authors concluded that the fate of oil in water is strongly directed by fast dissolution, adsorption, dispersion and agglomeration.

In an even more recent but contradicting article, Malinky and Shaw (143) have studied the association of specific petroleum conponents and suspended sediments and concluded that sedimentations via sorption to suspended mineral particles may not be a major pathway for the dispersion of petroleum in the marine environment. In their study, however, they used primarily glacially derived sediments

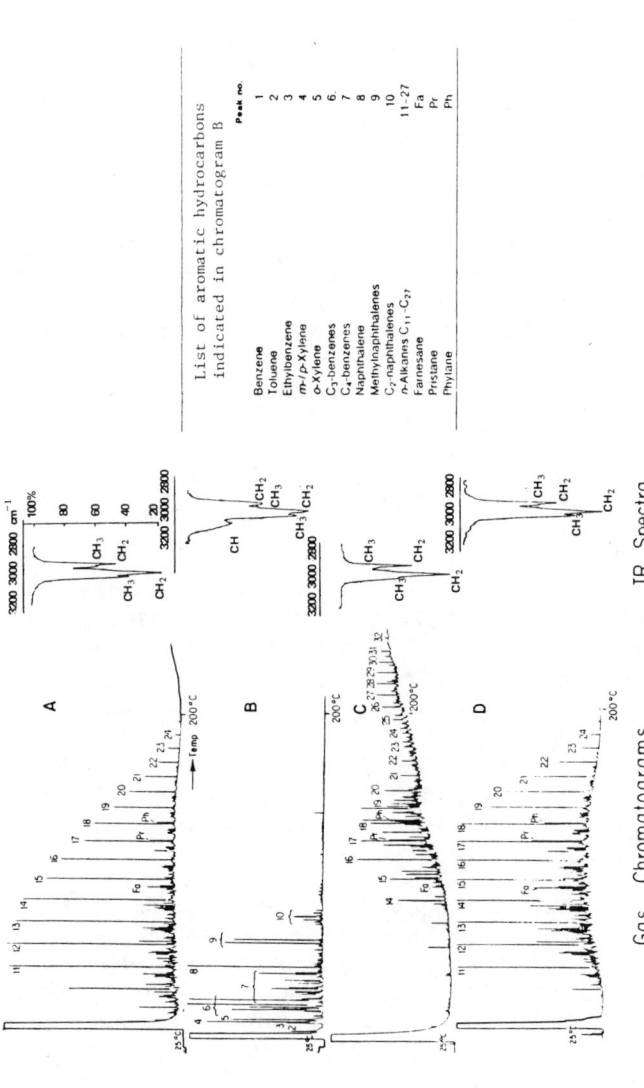

Figure I-22. Glass capillary FID gas chromatogram and partial IR spectra of extracts from oil-kaolinite modeling experiments (A = original oil, B = water-woluble oil fraction, C = dissolved oil adsorbed to kaolinite and D = oil-kaolinite agglomeration (from Zucher et al., 248). Reprinted with permission from Enviromental Science and Technology, Vol. 12, No. 7, © by the American Chemical Society.

from the south central Alaska region. These sediments are low in carbonates and organic carbon. Nevertheless, sediment from glacial erosion can occur in suspension in near-shore south central Alaskan marine waters in concentrations as high as 1 gram per liter. In their study, Malinky and Shaw (143) used 14C-labeled decane and biphenyl at near saturation levels. They found that in a saturated solution, the concentration of hydrocarbons associated with the sediments was approximately 30% of the original aqueous concentration in parts per million. From loading of permitted discharges into Port Valdez and measured sediment loads, the authors calculated that less than 3% of the oil released into the harbor could be associated with the sediment. Their model is valid only for oil in solution, however, so the mechanical coating of droplets (which Zürcher et al. [248] demonstrated to be a more efficient process) was not considered. Furthermore, Malinky and Shaw did not use a natural oil or even a water accommodated fraction of a natural oil, so the applicability of their work does suffer from some shortcomings. Nevertheless, their results led the authors to conclude that the adsorption to suspended particulate material is not significant and that the role of suspended mineral particulate material may be far less significant in adsorption of PAH (polycyclic aromatic hydrocarbons) in natural waters, than is the role of total suspended matter.

In addition to the simple adsorption and agglomeration from the physical contact of oil and suspended particulates as described above, Bassin and Ichiye (18) have demonstrated that oils and clays can form a spontaneous flocculation of colloids or colloidal electrolytes in the presence of dissolved salts. Their experiments suggest that oil sedimentation is caused mainly by adsorption of oil films onto clay particles which then subsequently flocculate due to electrolyte interaction, rather than by adsorption of discrete oil globules onto clay floccules.

Although Bassin and Ichiye used two types of natural oils, a Bunker C and South Louisiana crude, detailed results were presented only on the interaction of South Louisiana crude and the marine clay. The natural specific gravity of Bunker C was high enough that it sank spontaneously upon coagulation, even in the presence of the clays.

In general, flocculation behavior of hydrophobic materials such as clays is a function of a valence between the repulsive electrostatic fields of the negatively charged clay flakes and the attractive van der Waals forces between them. In a stable clay solution the repulsive force predominates; however, if a small amount of electrolyte is added to the solution, the electrical repulsion decreases and the van der Waals attractive forces can cause the particles to stick or flocculate (18). Further, while the organic structure of oils in colloidal micelles is not clearly understood (18), emulsified oil may be considered to be a colloidal suspension of oil micelles. These micelles also have specific charges associated with them, due to peptizing ions absorbed onto the micelle surface. Also, certain oils may contain oxygenated polar moieties which contribute to the net charge. In the presence of electrolytes the oil micelles can be induced to agglomerate in the same manner as clays, except that of instead of forming flocks, the oil molecules link up to form films. Thus, in an agglomerating dispersion containing both oil and clay, much of the oil electrolyte may be adsorbed onto the clay flakes by electrostatic bonding and this can induce clay flocculation by acting to shrink the double layer clouds.

Order of magnitude estimations on the amount of oil adsorbed by clay suggests that roughly 0.3-4 $cm^3$ of oil may adsorb per gram of clay under ideal conditions of salinity and clay dispersion. Thus, using the density of South Louisiana crude, Bassin and Ichiye (18) estimated that between 250 mg and 3 grams of oil may be accounted for by adsorption per gram of clay. Sedimented oil in excess of that capacity would be due to the incorporation of wet globules of oil into the clay/oil flocculates, and the presence of such oil globules was observed by microscopic examination. Interestingly enough, it was also observed that high salinity decreased the amount of oil adsorbed onto the clay as a result of the rapid formation of oil films and rapid flocculation of the individual clay particles to their mutual exclusion. Thus, depending on the type of oil and the relative concentration of oil and clay, there appears to be a threshold salinity above which there is a decreased adsorption of oils onto clays. Another striking aspect of the sedimented oil-clay agglomerates was their very coarse structure and increased porosity. Evidently, the adsorbed oil globules increases the spaces between clay flakes, thus affecting the size of the coagulated floccules. Also, since the oiled clays are lighter, their residence time and suspension are increased, thereby allowing a higher degree of flocculation and larger floccules. The authors concluded that although additional work is required to study the organic-rich, low sediment concentrations found in a typical dynamic environment, that the observed sedimention of oils by clay in coastal areas is due more to the electrolytic flocculation properties of the seawater than to any affinity between the oil and the clays themselves (18).

Perhaps one of the most exciting aspects of these oil particulate/sedimentation interaction studies is the agreement between the laboratory results showing the preferential partitioning of hydrocarbons by molecular weight with the results of deLappe et al. (61) and the results of Parker et al. (185) showing similar partitioning in the field. As noted by the studies of Malinky and Shaw (143), the nature of any suspended particulate matter is critical in its oil sorption properties and must be considered when predicting fates of oil in the water column.

Ingestion By Organisms

Another mechanism for the sedimentation of petroleum comes from the incorporation or ingestion of hydrocarbons by zooplankton, followed by the elimination of the petroleum in the feces. Following the Arrow incident in Chadebucto Bay, zooplankton were observed to ingest significant quantities of the Bunker C fuel oil droplets (51). In this instance most of the oil was eliminated in the feces which were observed to contain up to 7% oil by weight. From estimates made by Conover after the Arrow spill, it appears that zooplankton could remove as much as 20% of the oil droplets which were less than 0.1 millimeter in diameter from the water column. Because the density of the resultant copepod fecal matter was greater than that of the seawater, the encapsulated oil then rapidly sank. In a related study, Parker (183) found oil droplets in the gut content of copepods and barnacle larvae and in their fecal pellets. Evidently, as in the case of the copepods from the Arrow incident, the oil passed through the organisms chemically unaltered and the plankton did not appear to be adversely affected by the droplets. Long term impact studies have not been conducted to verify this conclusion, however, so some caution is required. Parker, Freegarde and Hatchard (184) estimated that one copepod (Calanus finmarchicus) could conceivably ingest up to $1.5 \times 10^{-4}$ grams of oil per day. A population of 2,000 individuals per cubic meter of seawater ingesting oil at this rate and covering an area of one square kilometer to a depth of ten meters, could therefore, theoretically, remove as much as three tons of oil daily if the oil concentration were 1.5 ppb or greater. Once removed, the oil incorporated into the fecal material would be more rapidly sedimented where it would then be either buried or subjected to further reworking.

## II. MICROBIAL DEGRADATION

INTRODUCTION

The fate of a petroleum spill in a marine/estuarine environment, as related to biodegradative processes, will encompass degradation, via microbial metabolism, ingestion by zooplankton, uptake and possible retention by marine invertebrates and vertebrates, as well as bioturbative effects. All serve to partition the petroleum hydrocarbons into the water column, biomass, and sediment regime of the ecosystem. However, due to both the immediate scope of this study and the constraints of time, attention will be directed towards an understanding of the influence of the indigenous microbial populations, in relation to the fate of a petroleum spill.

DISTRIBUTION AND TYPES OF HYDROCARBONOCLASTIC MICROORGANISMS

Various types of microorganisms capable of oxidizing petroleum hydrocarbons and related compounds (directly or by co-oxidation) are widespread in nature. Of particular interest in this review are those species occurring in the colder marine environments. Over 200 species of bacteria, yeasts, and filamentous fungi have been shown to metabolize one or more hydrocarbon compounds, ranging in complexity from methane to compounds of over 40 carbon atoms (247). This estimate will certainly prove to be conservative, both quantitatively and qualitatively, as even an achlorophyllus alga (<u>Protothea zophi</u>) has been shown to degrade petroleum and petroleum products (229, 230).

As would be expected from overall oceanic temperature distributions, the marine bacteria are predominantly psychrophilic (obligate and facultative) and show wide distributions even in the extremely cold Antarctic and Arctic environments. In the Arctic, petroleum degrading microorganisms are widely distributed in the Beaufort and Chukchi seas (9, 118), and relatively high population counts have been obtaineed from Arctic coastal waters (108). Even though these organisms may represent only small percentages of total heterotrophic microbial populations, studies have shown that populations can be as concentrated as those from more temperate oceans (118).

Psychrophilic pseudomonads (<u>Pseudomonas</u> sp.) are both ubiquitous and generally dominant species in marine environments (120). Other representative groups in the water column and sediments are <u>Arthrobacter</u>, <u>Cornybacterium</u>, <u>Vibrio</u>, <u>Enterobacter</u>, <u>Achromobacter</u>, <u>Brevibacterium</u>, <u>Aeromonas</u>, and <u>Acinetobacter</u> sp. (56, 167, 174, 204, 228).

Yeasts and fungi also have wide distributions in marine environments, including polar regions (120). <u>Cladosporium resinae</u> represents one of the most ubiquitous and prominent hydrocarbon-utilizing filamentous fungi (53, 227). The results of one study on other filamentous fungal species (186) suggest greater efficiency for consumption of hydrocarbon compounds relative to some bacteria or yeasts, perhaps in part due to formation of a "mat" over the surface of the petroleum tested. Table II-1 lists some of the microorganisms capable of degrading petroleum hydrocarbons and/or derived products.

The petroleum-degrading potential (capacity) for any mixed microbial population seems to be a function of both total numbers and types of hydrocarbonoclastic microorganisms present in the environment, as well as seasonal population fluxes. In the Beaufort Sea microbial populations are generally lower in winter than summer months with exception to sediment populations (118), and petroleum degrading microorganisms are present in ice at significantly lower numbers than in marine sediment or the water column (9). Of interest is an apparent response of hydrocarbonoclastic microorganisms to inputs of petroleums into their immediate environment. Microbes isolated from water and sediments from areas chronically exposed to oil pollution generally show greater capacities for degrading hydrocarbon class types (i.e. increased utilization of aromatic compounds) relative to populations from non-polluted areas (65, 105, 228, 234). Inputs of petroleum into the marine environment result in selection towards hydrocarbonoclastic organisms, causing a shift within the general population towards a relative increase in hydrocarbon-utilizing microbes (108). Areas suffering chronic oil pollution tend to have higher percentages of hydrocarbonoclastic microbes in the total heterotrophic population relative to less or nonpolluted areas (65, 132, 174), and in some cases these microbes may represent up to 100% of the total heterotrophic microbial population (167). Within a chronically contaminated area, the hydrocarbonoclastic microbial population in the water column may show greater capacity to utilize petroleum components, and be present in greater numbers, than the sediment population (228, 231). This may be in part due to greater and/or more consistent exposure to petroleum pollutants, as compared to the fluxes of these pollutants into the sediment environment through the water-sediment interface.

Table II-1. Microorganisms capable of oxidizing/cooxidizing petroleum hydrocarbons and/or their derivatives.

| Organism Type | Species Name | Source Environment(s)[a] | Reference(s) |
|---|---|---|---|
| Bacterium | Achromobacter sp. | T,M | 44,56,167 |
| | A. cycloclastes | (T,M)[b] | 191 |
| | Acinetobacter sp. | T,F,M,MS | 92,228,231 |
| | Aeromonas sp. | M,MS | 228 |
| | Alcaligenes sp. | | 68 |
| | A. eutrophus | | 92 |
| | Arthrobacter sp. | T,M | 56,123,193 |
| | Bacillus naphthalinicum | | 92 |
| | Beijerinekia sp. | | 90 |
| | Brevibacterium sp. | T,M | 56 |
| | B. healii | (T,M) | 68 |
| | Cellulomonas galba | | 68 |
| | Cornybacterium sp. | T,M | 56,120,191, 224 |
| | Flavobacterium sp. | FS,M | 80,92,105, 167,224 |
| | Micrococcus cerificans | | 69,211 |
| | Mycobacterium sp. | M,MS | 58 |
| | M. rhodochrous | | 224 |
| | Nocardia sp. | M,MS | 58,91,92,167, 169,191,224 |
| | N. coeliaca | (M,MS) | 191 |
| | N. coralina | (M,MS) | 113,114,190, 191 |
| | N. minima | (M,MS) | 190 |
| | N. opaca | (M,MS) | 80 |
| | N. salmonicolor | (M,MS) | 190 |
| | Pseudomonas sp. | T,F,M,MS | 5,8,56,91,92, 157,224,228, 231 |
| | P. aeruginosa | (T,F,M,MS) | 58,80,101,224 |
| | P. desmolytica | FS(T,F,M,MS) | 135 |
| | P. desmolyticum | (T,F,M,MS) | 92 |
| | P. Fluorescens | (T,F,M,MS) | 92 |
| | P. ligustri | (T,F,M,MS) | 68 |
| | P. methanica | (T,F,M,MS) | 130 |
| | P. oleovorans | (T,F,M,MS) | 13 |
| | P. orvilla | (T,F,M,MS) | 68 |
| | P. pseudomaleii | (T,F,M,MS) | 68 |
| | P. putida | FS,(T,F,M,MS) | 88,89,92,105 |
| | P. testosterni | (T,F,M,MS) | 92 |
| | Serratia marinoruba | M | 99 |
| | Streptomyces sp. | | 191 |

Table II-1, continued

| Organism Type | Species Name | Environment(s)[a] | Reference(s) |
|---|---|---|---|
| | Vibrio sp. | T,M,MS | 56,167,224,228 |
| Yeast | Candida petrophilium | (F,M) | 101 |
| | C. trophicalis | M,F | 52 |
| | Endomycopsis lipo-lytica | M,F | 52 |
| Fungi, filamentous | Aspergillus versicolor | | 186 |
| | Cephalosporium acremonium | | 186 |
| | Cladosporium resinae | T,F,M | 53,227 |
| | Cunningham elegans | | 186 |
| | Penicillum zonatum | | 186 |
| | P. ochro-chlorens | | 186 |
| Algae | Prothotheca zophi | M | 229,230 |

[a] T = terrestrial sediment; M = marine water column; F = freshwater; MS = marine bottom sediment; FS = freshwater bottom sediment.

[b] Key letters in parentheses designate the genus being indigenous to the specified environment.

## PETROLEUM HYDROCARBONS KNOWN TO BE OXIDIZED AND RATES OF UTILIZATION

The vast majority of experimental effort to determine which components of petroleums are utilized by microbes and the corresponding rates of utilization has been performed in the laboratory. Relatively little work has been done with controlled ecosystem enclosures or with attempts in the laboratory to simulate actual environmental conditions.

Utilization and rate studies have focused on two approaches: that of monitoring disappearance of individual petroleum hydrocarbon substrates introduced either alone or in mixtures to the microbial culture(s); or that of monitoring hydrocarbon disappearance (individual compounds or class types) from "whole" petroleums (crudes or refined products) used as growth substrates. The pure or mixed microbial cultures (bacteria, yeasts and fungi) have been obtained either by isolating strains from sediment or water samples, or by using whole sediment or water as an inoculum of the mixed naturally occurring microbial species. Isolation of pure or mixed cultures has been accomplished either by enrichment techniques (i.e. using a petroleum product as growth substrate) or by direct plating onto a nutrient medium.

Various techniques have been used to determine preferential utilization and the associated rates for petroleum hydrocarbons. Oxygen utilization and $CO_2$ evolution (as determined by manometric or specific analytical techniques), gas chromatography, and $^{14}C$-labeling, with the subsequent evolution of $^{14}CO_2$ detected by scintillation, have been used to monitor the degradation of petroleum hydrocarbon substrates in pure form or in mixtures. Radioactive carbon labeling has also been used to monitor degradation of $^{14}C$-labeled polycyclic aromatic hydrocarbons via high-performance liquid chromatography utilizing a fluorescent detector. The monitoring of degradation of whole petroleums has been accomplished with gas chromatography, combined gas chromatography-mass spectrometry, and computerized mass spectrophotometry.

Table II-2 lists some of the petroleum hydrocarbon compounds known to be oxidized (directly or by co-oxidation) by hydrocarbonoclastic microorganisms. As with the list of hydrocarbon-utilizing organisms, this list will surely prove to be conservative with further studies.

Results from preferential utilization and rate studies complicate the generally accepted scenario of microbial petroleum degradation (i.e. n-alkanes being oxidized more

Table II-2. Petroleum hydrocarbon compounds known to be oxidized (directly or co-oxidized) by hydrocarbonoclastic microorganisms.

| Class Type | Compound | Reference(s) |
|---|---|---|
| n-Alkanes | Ethane | 80,130,186 |
| | Propane | 80,186 |
| | Butane | 85,186 |
| | Pentane | 186 |
| | Hexane | 186,224 |
| | Heptane | 186,224 |
| | Octane | 186 |
| | Decane | 80,135,186,224, 228,232 |
| | Undecane | 186,228,232 |
| | Dodecane | 80,186,197,212, 227,228,232 |
| | Tridecane | 186,228,232 |
| | Tetradecane | 80,186,212,228, 232 |
| | Pentadecane | 186,228,232 |
| | Hexadecane | 80,91,123,131, 167,186,211,227, 228,232,234 |
| | Heptadecane | 131,186,228,232 |
| | Octadecane | 131,180,212,224, 228,232 |
| | Nonadecane | 186,228,232 |
| | Eicosane | 228,232 |
| | N-$C_{14}$ to N-$C_{30}$ | 119 |
| Alkenes | $C_3$ to $C_{11}$ | 186 |
| | Hexadecene-1 | 167 |
| Branched Alkanes | Branched to 12 carbons | 186 |
| | 2-Methylhexane | 224 |
| | Pristane | 120,167,186,209, 228,232 |
| | Phytane | 119,209 |
| | Farnesene | 209 |
| Cyclic Alkanes | 1-Ring (General Class) | 231,233,235 |
| | Cyclohexane | 80,228,234 |
| | n-Butylcyclohexane | 58,228,234 |
| | Heptacyclohexane | 209 |
| | 2-Ring (General Class) | 231,233,235 |
| | 3-Ring (General Class) | 231,233,235 |
| | 4-Ring (General Class) | 231,233 |
| | 5-Ring (General Class) | 231,233 |
| | 6-Ring (General Class) | 231 |

Table II-2, continued

| Class Type | Compound | Reference(s) |
|---|---|---|
| Aromatics[a] | | |
| Monoaromatics | Benzene(s) | 231,233 |
| | Alkylbenzenes (Linear/Branched) | 88,92,131,133, 191,224 |
| | Cycloalkylbenzenes | 92 |
| | Ethylbenzenes | 58,88,114,224 |
| | Trimethylbenzenes | 92,114 |
| | Tetramethylbenzenes | 92 |
| | n-Butylbenzene | 113 |
| | n-Propylbenzend | 58,224 |
| | Chlorobenzene | 92 |
| | n-Amylbenzene | 67 |
| | Naphthenebenzene(s) | 231 |
| | Dinaphthenebenzenes | 231 |
| | Benzcycloparaffins | 235 |
| | Benzdicycloparaffins | 235 |
| | p-Cresol | 80 |
| | Cumene | 191 |
| | Pseudocumene | 228,232 |
| | 3-Phenyleicosane | 80 |
| | Toluene | 88,92,120,131, 224,234 |
| | halo-Toluenes (Cl,Br,F) | 92 |
| | p-Isopropyltoluene | 224 |
| | Tetralin | 191 |
| | Xylenes (para-,meta-) | 89,91,92,113, 120,190,191 |
| Diaromatics | General Class | 80,88,91,92,99, 105,107,120,131, 191,205,224,228, 231,232,234,235 |
| | Naphthalene | 234,235 |
| | Methylnaphthalenes (mono-, di) | 80,92,120,131, 133,190,224 |
| | Chloronaphthalenes (1-,2-) | 80,92 |
| | 1-Bromonaphthalene | 80 |
| | Acenaphthenes-dibenzofurans | 231,236 |
| | 1-(2-Naphthyl) undecane | 80 |
| | Decalin (decahydronaphthalene) | 167,209 |
| | Biphenyl | 88,90,92,191,209 |
| Triaromatics | General Class | 105,233 |
| | Anthracene(s) | 80,90,92,105, 107,133,167,205, 224 |

Table II-2, continued

| Class Type | Compound | Reference(s) |
|---|---|---|
| | Phenanthrene(s) | 80,88,90,92,133, 205,224,228,231, 232,235 |
| | Methylphenanthrene(s) | 224 |
| | Naphthenephenanthrenes | 231 |
| | Sterane(s) | 203 |
| Tetraaromatics | General Class | 233 |
| | 1,2-Benzanthracene (Benz(a)anthracene) | 120,205,228,232 |
| | Benzanthracene(s) | 247 |
| | Benzo(a)anthracene | 92,105,235 |
| | Chrysenes | 231 |
| | Pyrene(s) | 92,228,231,232 |
| | 20-Methylcholanthrene | 247 |
| | Terpanes | 203 |
| Pentaaromatics | General Class | 231 |
| | Dibenzanthracene(s) | 228 |
| | 1,2,5,6-Dibenzanthracene | 120,205 |
| | Benzo(a)pyrene (3,4-Benzpyrene) | 90,92,105,120, 136 |
| | Perylene | 228,232 |
| | Benzperylene | 92 |
| | Hopanes | 203 |
| Sulfur Aromatics | General Class | 233 |
| | Benzothiophene(s) | 231 |
| | Dibenzothiophene | 120,167 |

[a] Aromatic class types are defined by the total number of aromatic rings present in the particular compound.

readily than iso-alkanes, cyclic hydrocarbons, or olefins). Data from studies monitoring degradation for whole petroleums by sediment bacteria in a temperate marine/estuarine environment (235), and by water and sediment bacteria in the Arctic marine environment (9, 108) indicate both even or similar rates of utilization and simultaneous degradation for various petroleum hydrocarbon class types and individual components. Different results were obtained from a kinetic study (233) which assessed whole petroleum degradation. Total oil residue decreased exponentially with time with the maximum decrease at the logarithmic phase of bacterial growth (with concomitant increase in proportions of asphaltenes and resins). The saturate fraction decreased continuously with time. Results indicated a dynamic process in which the petroleum components are degraded simultaneously and at different rates.

However, the majority of studies have provided data supporting the scenario of preferential utilization with differential rates between hydrocarbon compounds and class types. Studies using mixed cultures of both marine and terrestrial bacteria have shown preferential utilization for normal and branched paraffins (119, 136, 160, 247) over the higher molecular weight fractions. Oxidation rates have been shown to be inversely proportional to chain length, with isoprenoids (pristane and phytane) being utilized only after the n-paraffins (119), although isoprenoids have also been shown to be oxidized simultaneously (160). The relatively refractory isoprenoid pristane can be oxidized, but by fewer hydrocarbonoclastic species than phytane or farnesane (209). Other studies do not delineate any consistent pattern(s) for utilization of other class types, such as cyclic alkanes or aromatic compounds (131, 167, 209, 234). To accurately assess preferential utilization it is important to consider both compound specificity by a particular microbe as well as the numbers and types of compounds the microbe is able to oxidize. Studies have shown that bacteria and fungi isolated and grown on substrates containing relatively complex hydrocarbons (i.e. aromatics) can utilize a wider range of class types than microbes isolated from less complex hydrocarbon substrates, such as normal or branched alkanes (186, 222).

Rather complex polycyclic aromatic hydrocarbons are oxidized by marine microbes resulting in short residence times for aromatics in marine waters (133). Some polycyclic aromatics are more refractory, such as steranes, terpanes, and hopanes (203) as well as carcinogenic aromatics such as benzo($\alpha$) pyrene and benzo($\alpha$)anthracene (105, 131, 205, 232, 247). Such aromatics reaching the deep-sea environment may still be subject to microbial degradation, although it is believed that rates of utilization would be much slower relative to surface or mid-depth environments (11, 116, 120).

## FACTORS AFFECTING UTILIZATION AND RATES

### Nutrients

The importance of nutrient concentrations to microbial degradation of petroleums and individual petroleum hydrocarbons has been well documented in both laboratory and field studies. Nitrogen and phosphorous (available as $NO_2$, $NO_3$, $NH_4$ and $PO_4$) have been shown to be limiting factors to both rates and extents of petroleum compound degradation (9, 17, 52, 120, 169, 193, 232), as well as having a stimulating effect by addition of nutrient supplements (e.g. $[NH_4]_2SO_4$ and $K_2HPO_4$) to the immediate experimental environment (65, 108, 193).

The effect of iron on petroleum degradation in seawater has been studied (65) and results suggested that it may become limiting when precipitated out of the environment as ferric hydroxide, under alkaline conditions. However, both the natural abundance of iron in the lithosphere and marine pH ranges would probably prevent this limitation from occurring.

### Temperature

The environmental temperature can affect degradation rates by acting upon the microbial populations in several ways. Ambient temperatures, whether in the laboratory or in natural conditions, will select for microbial species tolerant to the temperature range present, such as psychrophilic bacteria with optimal growth rates from $15°-20°C$ (120). Thus, qualitative shifts may occur within the microbial population (and in the inherent petroleum degradative capacity), as reflected by the relative presence of hydrocarbonoclastic microbes. Low temperatures generally suppress degradation rates (9, 17, 245, 247) by suppressing growth rates and metabolic rates of the microbes involved (169) and/or by actually inhibiting growth due to increased retention of toxic components in the petroleum. Inhibition due to toxic volatile compounds (5, 8, 17) that evaporate more slowly at low temperatures, or to the increased solubilities of potentially toxic petroleum compounds at higher temperatures (228) may occur.

## Salinity

Changes in environmental salinity may result in qualitative and quantitative shifts in microbial populations, as with changes in ambient temperatures. Many freshwater and estuarine microbes can survive higher salinities, although few reproduce. Conversely, most marine species have an optimum salinity range of 25-35% and reproduce poorly, or not at all, in salinities from 15-20% (247). One laboratory study using pure bacterial cultures demonstrated inhibition of emulsification of a crude oil by high salt concentrations (245). Such drastic variation in salinities most likely would occur only in estuarine environments, and it appears that salinity is most important to overall oil degradation in terms of its influence on compound solubility (as discussed elsewhere).

## Oxygen

Both free and dissolved oxygen levels in the immediate environment are important to microbial degradation (120), as dictated by both microbial growth and the oxygen requirements for complete oxidation of petroleum hydrocarbons. Approximately 3 to 4 mg of molecular oxygen per mg hydrocarbon in required for complete conversion to $CO_2$ and water, although less oxygen is required if some of the oxidized intermediates are converted into cell biomass (247).

## Presence of Ice

Relatively little is known about the effects of ice presence upon microbial populations. In the Beaufort Sea, petroleum degrading microorganisms are present in ice only in low numbers, relative to local marine sediment and water (9). One study in this area demonstrated a negative effect by ice presence on the underlying water microbial population (118).

## INTERFERENCE/ENHANCEMENT OF DEGRADATION RATES

### Interference with Chemotaxis

The response of hydrocarbonoclastic microbes to environmental input of petroleums may be inhibited if such petroleums contain compounds that interfere with the chemotaxis of the organisms. A wide variety of chemical and physical water pollutants including benzene, kerosene, a Kuwait crude oil and other substances (24, 197, 236) inhibit microbial chemotaxis by reversible blockage of chemoreceptors. However, acclimation to petroleum compounds may decrease the magnitude of this interference (24).

## Toxicities to Hydrocarbonoclastic Organisms

As discussed previously, ambient temperatures may have negative effects on petroleum-degrading microbes due to retention of toxic volatile compounds at low temperature (5, 8, 17), or to increased solubility of toxic components at higher temperatures (228).

Certain petroleum components may be bacteriostatic or bacteriocidal, such as phenol, toluene, and cyclohexane. However, when oil is mixed with seawater, the concentrations of these components will not be high enough to be toxic and they can actually then be metabolized by the indigenous microbes (7).

Addition of emulsification agents to an oil spill may indirectly affect microbes by way of influencing dissolution rates for petroleum components. Oil-in-water emulsification will increase the surface area of the spill film, thereby decreasing the diffusion time required for the more water-soluble components to leach into the water column. The residue then may become more toxic due to retention of relatively toxic and less soluble naphthenic and aromatic compounds (49).

The concentration of dissolved hydrocarbons may produce toxic effects at high or saturation levels for specific petroleum components (247). Saturation levels for hydrocarbons in the $C_5$ to $C_8$ range can inhibit microbes by interfering with membrane transport functions (197), such as phosphate transport.

As a petroleum spill undergoes weathering, oxidized products form which may represent increased toxicity over the "parent" compound. One example is dihydroxynaphthalene (a partial oxidation product of naphthalene) which is more toxic to some marine bacteria than its precursor (99).

Metabolic products by the hydrocarbonoclastic microbes themselves may also be more toxic than the parent compound. Aerobic degradation, with incorporation of molecular oxygen, can result in intermediates with greater solubility and toxicity. An example would be the conversion of butylbenzene to 3-phenylpropionic acid by a Pseudomonas sp. (49). Certain yeasts have been shown to increase the toxicity of crude oil toward the common guppy (Lesbistes reticulatas), possibly due to production of metabolic products by the yeasts degrading the oil (52).

One more mechanism of toxicity should be considered, which concerns the rapid change in the microenvironment of the microbes involved. Upon contact with oil droplets, microorganisms tend to coalesce. This decreases mobility and, consequently, the rate of oil slick inoculation for the biodegradation process (197).

## Effects of Chemical Dispersants

Several studies have been performed to assess enhancement of microbial degradation by various chemical dispersants (surfactants) using mixed cultures of terrestrial, marine and freshwater bacteria (168, 195, 196). The majority of dispersants tested showed enhancement of biodegradation by emulsification of the petroleums examined. Modes of enhancement seem to be due to the increased surface area (bacterial growth rates increase with interfacial area) and facilitation of microbial diffusion to the interfacial area between the oil film and aquatic environment (196). Potentially negative effects (as discussed previously for toxicities to organisms) may be related to effects on chemical uptake by the microorganisms due to action on cell membranes by surfactants.

## Microbially-produced Emulsification Factors

Hydrocarbonoclastic bacteria and yeasts produce biochemical intermediates and enzymes which can act as surface active agents (244). Emulsifiers have been shown to be produced from degradation of the paraffinic petroleum components, and were initially characterized as high molecular weight polysaccharides (245). A yeast (Candida petrophilium) and a bacterium (Pseudomonas aeruginosa) produce heat-stable emulsifying factors which are effective over a wide pH range in emulsifying hydrocarbon mixtures and crude oils (101).

## Enhancement of Degradation (Induced Stimulation)

Increased degradation (extent and rates) by addition of oleophilic fertilizers to the immediate spill environment has been demonstrated by studies utilizing pure and mixed marine bacteria cultures in both laboratory and field situations. Various water-soluble nitrate and phosphate salts, and an oleophilic fertilizer were compared for effectiveness in degrading a paraffinic crude. The greatest extent of degradation was obtained with the oleophilic fertilizer or $KNO_3$ (nitrate source) combined with octylphosphate (phosphate source). The oleophilic fertilizer selectively supplies nutrients to hydrocarbonoclastic microbes and does not trigger algal blooms as with the water-soluble salts (6). Octylphosphate, in combination with paraffinized urea and ferric octoate, has been shown to stimulate crude oil degradation by marine microbes at favorable water temperatures (65). Paraffin-supported fertilizers containing $MgNH_4PO_4$ as an active ingredient have demonstrated increased microbial degradation rates for various crudes (177, 178), although paraffinized particulates may become agglomeration centers for relatively viscous oils, thus decreasing the surface area exposed to microbial attack. Nutrient release is controlled by physical factors (thickness and continuity of paraffin coat, water temperature, etc.) rather than by the nutrient "needs" of the microbes (178). Other oleophilic nutrient sources tested that demonstrated substantial increases in biodegradation rates are lecithin (choline phosphoglyceride) for phosphorous, combined with either ethyl allophanate or phenyl hydantoin for nitrogen. Results from degradation experiments for ethyl allophanate suggest the following mechanism (178):

$$\underset{\underset{O}{\|}\phantom{xx}\underset{O}{\|}}{Et-O-C-NH-C-NH_2} \rightarrow Et-OH + \underset{\underset{O}{\|}\phantom{xx}\underset{O}{\|}}{HO-C-NH-C-NH_2} \rightarrow CO_2 + \underset{\underset{O}{\|}}{NH_2-C-NH_2}$$
$$\text{(urea)}$$

## METABOLIC PATHWAYS OF MICROBIAL PETROLEUM DEGRADATION

Laboratory studies attempting to determine specific metabolic pathways have largely utilized pure bacterial cultures, which may or may not be represented in the marine environment. Each species of hydrocarbonoclastic microbes generally utilizes only a narrow range of homologous petroleum hydrocarbons (247), and variations in proposed pathways for a particular compound have resulted from studies using the same or very similar strains. However, it is fortunate for the purpose of this review that the microbes best studied are ubiquitous in marine environments at the genus level (e.g. <u>Pseudomonas</u>).

The enzymes involved in the metabolic pathways are both constitutive and adaptive/inducible in nature. Products of biodegradation include $CO_2$, water, various hydroperoxides, alcohols, phenols, carbonyls, aldehydes, ketones, fatty acids and esters (191, 224, 247). Complete oxidation (to $CO_2$) of petroleums is facilitated by mixed microbial cultures. Hydrocarbon components utilized for growth can be converted/incorporated into microbial biomass as proteins, amino acids, nucleic acids, purines, pyrimidines, lipids, and polysaccharides (247). Microbial oxidation of hydrocarbon compounds and their derivatives may proceed via direct oxidation or co-oxidation, providing a suitable growth substrate is available. Generally, hydrocarbon co-oxidation reactions involve the incorporation of molecular oxygen by mono- and dioxygenases (191). Table II-3 indicates some of the petroleum hydrocarbons degraded by co-oxidation, the growth substrate(s), and the microbial species involved.

## Alkanes

Studies utilizing bacteria, yeasts, and fungi have proposed oxidation modes for paraffins involving monoterminal oxidation (13, 53, 80, 135, 191, 224, 227). The general scenario for oxidation of medium chain length alkanes (i.e. $C_6 - C_{12}$) involves initial oxidative attack at the terminal methyl group, forming the primary alcohol. This is sequenced by aldehyde formation, followed by conversion to the corresponding fatty acid. Studies with two strains of a filamentous fungi (Clodosporium resinae) suggest the following pathway for monoterminal oxidation (53, 227):

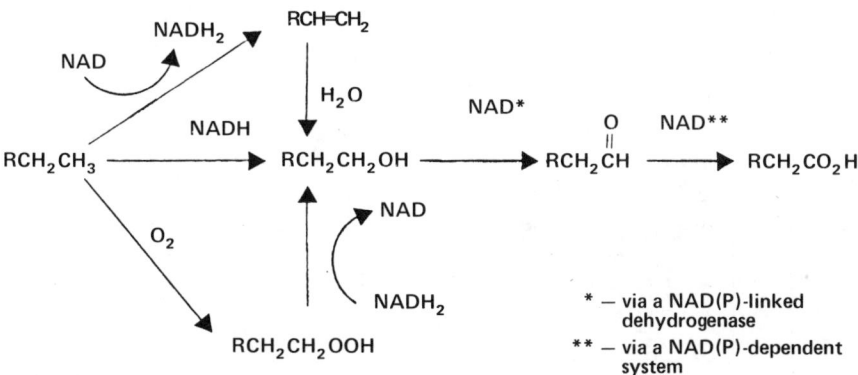

The fatty acid formed in this pathway may be further oxidized by β-oxidation (224, 227) or may be directly incorporated into cellular lipid material. However, the study with C. resinae (227) showed no correlation between cellular fatty acids and n-alkane growth substrates.

The involvement of nicotinamide adenine dinucleotide (NAD) in the initial steps of n-alkane monoterminal oxidation is further demonstrated by the mechanism for heptane degradation by a Pseudomonas sp., as postulated by Azoulay et al. (224):

Table II-3. Microbial co-oxidation of petroleum hydrocarbon compounds (reaction figures reprinted with permission from Lipids, Vol. 6, No. 7, © by the American Oil Chemists' Society).

| Growth Substrate | Co-oxidation Reaction | Bacterium | Reference(s) |
|---|---|---|---|
| Methane | Ethane → ethanol, acetaldehyde, acetic acid<br>Propane → n-propanol, propionic acid, acetone<br>n-Butane → n-butanol, n-butyric acid, 2-butanone | Pseudomonas | 80,130 |
| Hexadecane or pentadecane salts medium with yeast extract or corn steep liquor supplements | n-Hexadecane → n-hexadecanone | Arthrobacter sp. | 123 |
| n-Alkanes or an alkyl moiety | n-butylcyclohexane → cyclohexane acetic acid<br>ethylbenzene → phenylacetic acid<br>n-propylbenzene → phenylacrylic acid | Nocardia sp.<br>Myobacterium sp.<br>P. aeruginosa | 58 |

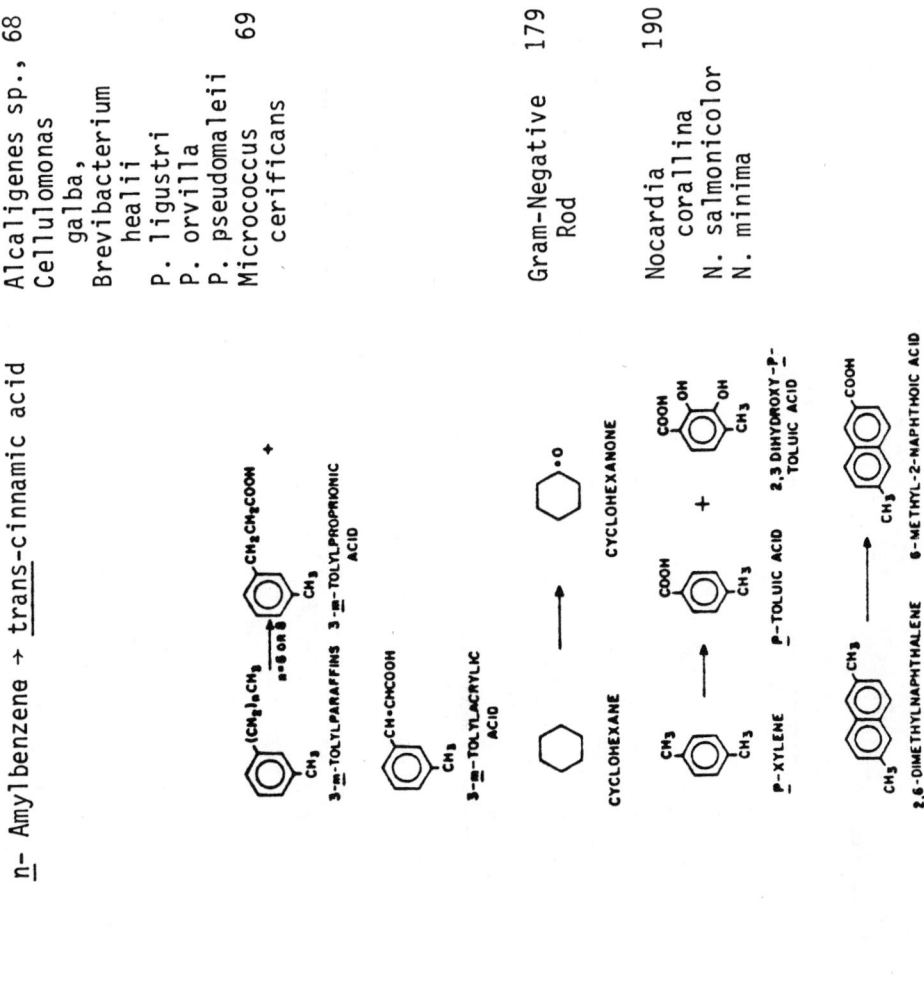

Table II-3, continued

| Growth Substrate | Co-oxidation Reaction | Bacterium | Reference(s) |
|---|---|---|---|
| n-Paraffins | p-XYLENE → 3,6-DIMETHYLPYROCATECHOL + α,α'-DIMETHYL-CIS,CIS-MUCONIC ACID | N. corallina | 113 |
| | PSEUDOCUMENE → 2,5-DIMETHYLBENZYL ALCOHOL + 3,4-DIMETHYLBENZOIC ACID + 2,3-DIHYDROXY-4,6-DIMETHYL BENZOIC ACID | N. corallina | 191 |
| | 1,2,3-TRIMETHYLBENZENE → 2,6-DIMETHYL BENZYL ALCOHOL + 2,3-DIMETHYL BENZOIC ACID | N. corallina | 114, 191 |
| | p-DIETHYLBENZENE → p-DIACETYLBENZENE + 4-(1-HYDROXYETHYL)-ACETOPHENONE | | |

## MICROBIAL DEGRADATION

| Substrate / Pathway | Organism | Ref. |
|---|---|---|
| n-Hexadecane or Cerelose: Biphenyl → 4-benzoyl-butyric acid; Tetralin → 4-phenyl-(2'-hydroxy)-butyric acid | | |
| Naphthalene → 4-(2-hydroxyphenyl)-2-ketobutyric acid | Nocardia sp.<br>N. coeliaca<br>Streptomyces sp. | 191 |
| Biphenyl → 4-benzoylbutyric acid<br>p-Xylene → dimethylmuconic acid<br>Benzene → phenol<br>Naphthalene → gentisic acid | Nocardia sp.<br>N. coeliaca | 191 |
| | Achromobacter cycloclastes | 191 |
| Naphthalene → salicyclic acid<br>Anthracene → 2-hydroxy-3-naphthoic acids | Corynebacterium sp. | 191 |
| Cerelose (will occur very slowly without cosubstrate) cosubstrate | | |

# 74  FATE AND WEATHERING OF PETROLEUM SPILLS

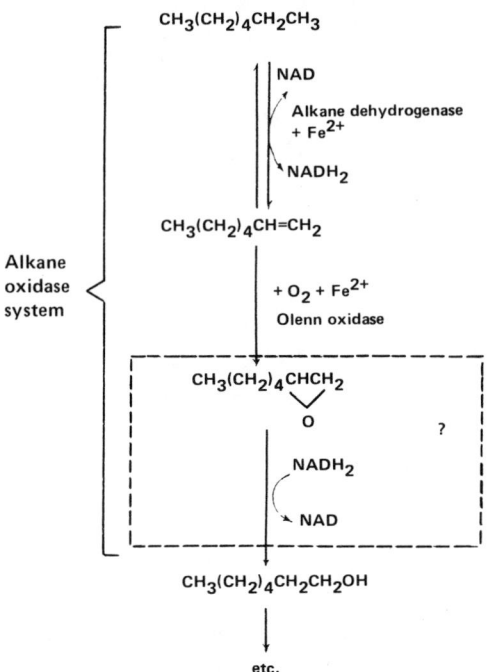

Other mechanisms possible include diterminal oxidation to the corresponding dioic acid, or subterminal oxidation resulting in formation of alkenes, secondary alcohols, and ketones. Examples of diterminal attack ($\alpha$, $\omega$-oxidation) have been provided by a <u>Corynebacterium</u> sp. for $C_{10}$-$C_{14}$ alkanes (224), as well as those listed in Table II-4 (eq. 5, 7, 9). Subterminal carbon oxidation of gaseous alkanes has been proposed for <u>Pseudomonas methanica</u> by Leadbetter and Foster (80) with the following pathway:

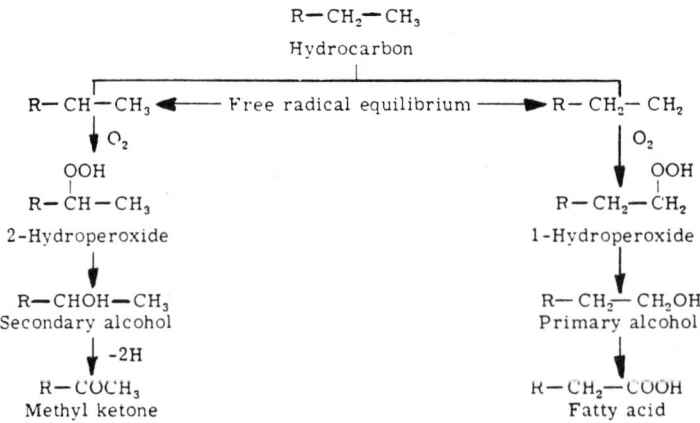

Table II-4. Examples of microbial n-alkane oxidations.

| | Oxidation Reaction | Microbial Species | Reference |
|---|---|---|---|
| (1) | Hexane → Hexanoic Acid | Pseudomonas aeruginosa | 224 |
| (2) | Heptane → Heptanoic Acid | P. Aeruginosa | 224 |
| (3) | Octane → n-Octanol ↔ Octaldehyde → Octanoic Acid | P. Olevorans | 13 |
| (4) | n-Decane → n-Decanol + n-Decanoic Acid | P. desmolytica | 135 |
| (5) | n-Decane → n-Decanedioic Acid | Gram-Positive Rod | 80 |
| (6) | Dodecane → Ester with Palmityl Moiety | Micrococcus cerificans | 44 |
| (7) | n-Dodecane → n-Dodecanedioic Acid | Gram-Positive Rod | 80 |
| (8) | Tetradecane → Tetraderyl (Myristyl) Palmitate | Micrococcus cerificans | 212 |
| (9) | n-Tetradecane → n-Tetradecanedioic Acid | Gram-Positive Rod | 80 |
| (10) | Hexadecane → Hexadecanoic Acid | Cladosporium Resinae | 227 |
| (11) | Hexadecane → Cetyl Palmitate | Gram-Negative Coccus | 211 |
| (12) | Octadecane → Octadecyl Stearate + Octadecyl Palmitate | Gram-Negative Coccus | 212 |

However, no methyl ketones were detected for n-alkanes with more than six carbons (80). Either metabolism of longer chains does not involve methyl ketones or they are transformed too quickly for accumulation.

Formation of esters has been noted for $C_{12}$-$C_{20}$ n-alkanes (211, 212, 224) by Micrococcus cerificans (Table II-4, Eq. 6, 8, 12). The following pathway for oxidation of n-hexadecane to Cetyl palmitate by M. cerificans has been proposed by Stewart et al. (80) from results of $^{18}O$ incorporation studies. It was suggested that the pathways may be similar for n-dodecane, n-tetradecane, and n-octadecane:

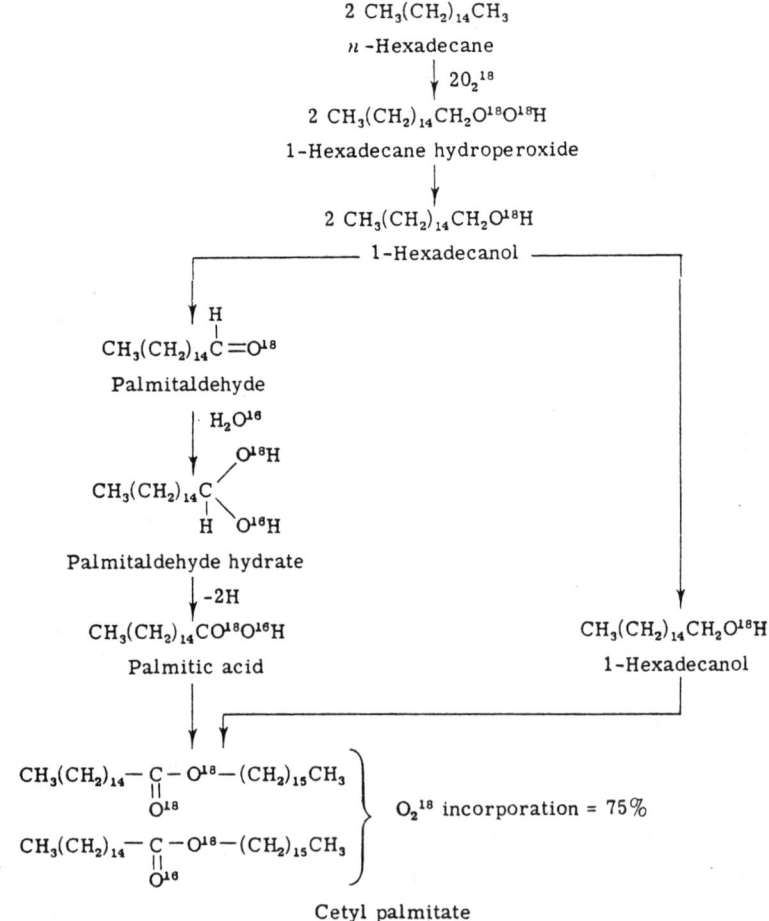

Cetyl palmitate
(2 Molecules of different isotopic composition)

(Reprinted with permission from Academic Press, copyright 1962.

Isoparaffin hydrocarbons show decreased or even negligible oxidation by microbes, relative to n-alkanes. However, 2-Methyl hexane has been shown to be oxidized by a Pseudomonas sp. with two probably pathways, although the $C_6$ pathway was found to be favored (224).

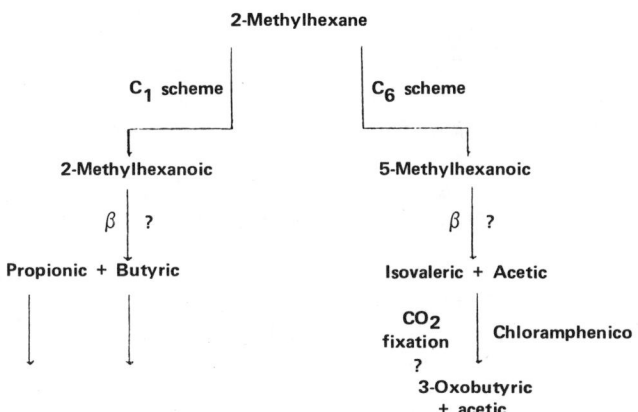

Cycloparaffins seem to be poorly utilizable by microbes in general, especially those with short alkyl substituents unsuitable for use as growth substrates (224). Very little information is available pertaining to microbial degradation, although dihydrodial formation (characteristic of aromatic degradation) may be plausible. Eliot et al. (80) proposed the following pathway for early stages of cyclohexane oxidation in the rabbit:

(Reprinted with permission of Academic Press, copyright 1962)

Olefins have been shown to be oxidized by bacteria and yeasts. One yeast investigated (Candida lipolytica) with α-olefins demonstrated initial oxidation of the double bond, with further degradation of the 1,2-diol formed (33). Bacteria initially attack α-olefins at the unsaturated end (224), although the terminal double bond may also be oxidized.

## Aromatic Hydrocarbons

Microbial oxidation of aromatic hydrocarbons has received greater attention than for alkanes, and aromatics have been shown to be oxidized directly and by co-oxidation. Table II-3 presents examples of aromatics that are oxidized if suitable growth cosubstrates are present.

Several studies have shown the presence of *cis*-dihydrodiol intermediates during aromatic oxidations (88, 89, 90, 91, 92, 132, 191, 224) for microbial species, although the *trans*-dihydrodiol is formed in mammalian systems (80, 132). these dihydrodiol intermediates are then converted to *ortho*-hydroxy derivatives called catechols. Cathechol formation has been proposed via the following mechanism (80, 92).

$$\text{benzene} \xrightarrow[\text{bacteria}]{O_2} \text{(cis-dihydro-peroxide)} \xrightarrow{2H^+ + 2e} \text{(cis-diol)} \xrightarrow[NADH_2]{NAD^+} \text{(catechol)}$$

Dihydrodiol intermediates also exist for aromatics with fused benzene rings, such as naphthalene and anthracene (92):

Naphthalene $\xrightarrow{\text{oxidation}}$ *cis*-1(R),2(S)-dihydroxy-1,2-dihydronaphthalene

Anthracene $\xrightarrow{\text{oxidation}}$ *cis*-1(R),2(S)-dihydroxy-1,2-dihydroanthracene

The formation of *cis*-dihydrodiol intermediates has been shown for several aromatic hydrocarbons by Pseudomonas sp. (101):

The catechol product(s) undergo further degradation by rupture of the benzene ring by both *ortho* and *meta* cleavage (92, 191), relative to the two hydroxyl groups whose presence is necessary for enzymatic fission (91, 92). Benzene ring oxidation by Mycobacterium rhodochrous and P. aeruginosa has been proposed to result in ring fission via the following pathway (144):

These intermediate products then can enter the tricarboxylic acid (TCA) system for further degradation (120).

The mechanisms on the following page have been proposed (224) for the conversion of Catechol to 3-Oxoadipic acid [A] and for conversion of Catechol to Acetic and Pyruvic acids [B].

A monohydroxylation scheme for conversion of Benzene to Muconic acid has been proposed for a Nocardia sp. (80):

(Reprinted with permission from Academic Press, copyright 1962.)

Several substituted monoaromatics have been studied in relation to bacterial degradation. Modes of oxidative attack center on either the side chain substituent or on ring fission (91, 120). Alkyl-side chain attack is similar to that for aliphatic compounds by $\beta$-oxidation. The acids thus formed (or precursors of) for odd-numbered side chains and even-numbered chains are benzoic and phenylacetic acids, respectively (120). A Pseudomonas sp. has been shown to convert butylbenzene to 3-phenylpropionic acid (49), which may represent an intermediate to the expected product, phenylacetic acid.

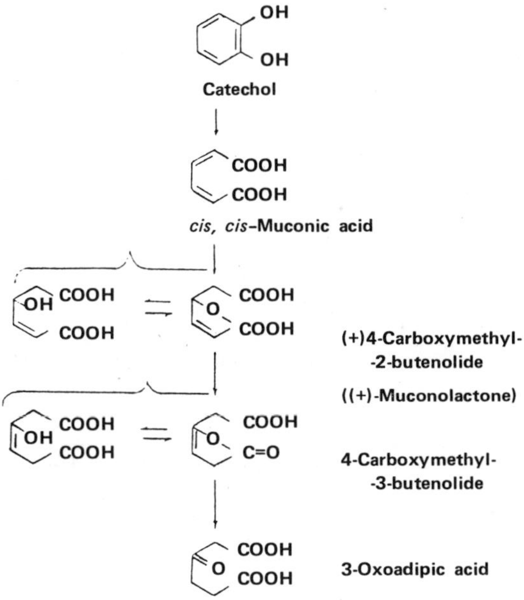

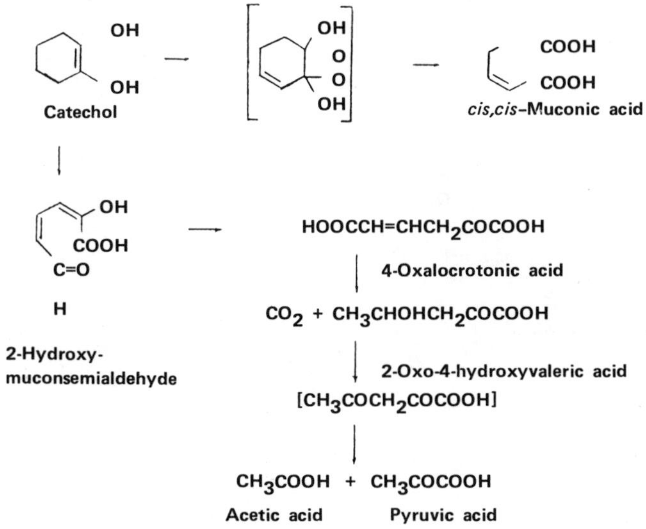

# MICROBIAL DEGRADATION

Catechol → cis,cis-Muconic acid

(+)4-Carboxymethyl-2-butenolide
((+)-Muconolactone)

4-Carboxymethyl-3-butenolide

3-Oxoadipic acid

Catechol → cis,cis-Muconic acid

2-Hydroxy-muconsemialdehyde → HOOCCH=CHCH$_2$COCOOH

4-Oxalocrotonic acid

$CO_2$ + CH$_3$CHOHCH$_2$COCOOH

2-Oxo-4-hydroxyvaleric acid
[CH$_3$COCH$_2$COCOOH]

CH$_3$COOH + CH$_3$COCOOH
Acetic acid  Pyruvic acid

(Reprinted with permission from Advances in Enzymology, Vol. 27, copyright 1965, John Wiley and Sons, Inc.)

Toluene and alkylbenzenes with odd-numbered side chains may be microbially degaded via the following proposed pathways (224):

Toluene ($CH_3$–Ph) → Benzoic acid (COOH–Ph) → [epoxide intermediate COOH] → [COOH, OH, H, OH cyclohexadiene] → Catechol (OH, OH)

Benzoic acid → Salicylic acid (COOH, OH) → Catechol

Alkylbenzene ($CH_2(CH_2)_{2n-1}CH_2$–Ph) → Phenylpropionic acid ($CH_2CH_2COOH$–Ph) ⇌ Cinnamic acid (CH=CHCOOH–Ph) → Benzoic acid

Phenylpropionic acid → n-Hydroxyphenyl-propionic acid ($CH_2CH_2COOH$, OH) → 2,3-Dihydroxyphenyl-propionic acid ($CH_2CH_2COOH$, OH, OH) → ?

Oxidation of ethylbenzene by a Nocardia sp. occurs (as a possible co-oxidation reaction) when n-octadecane or n-hexadecane is present with the following proposed sequence (224):

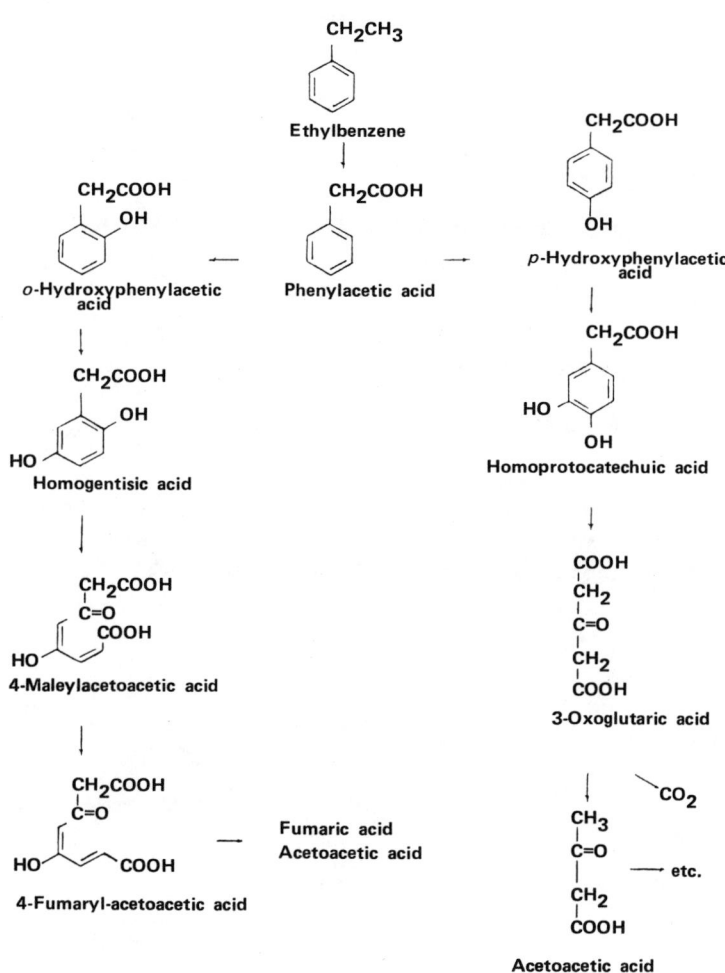

Preferential attack on alkyl substituents was also demonstrated by Norcardia opoca for two other even-numbered alkylaromatics by Webley et al. (80), and is believed to occur by terminal attack to the homologous fatty acid, followed by successive α-oxidations (80):

CH$_2$(CH$_2$)$_{8, 10,}$or $_{16}$CH$_3$ — Phenyl → CH$_2$COOH — Phenyl

*n*-Alkylbenzene          Phenylacetic acid

CH$_3$CH$_2$CH$_2$(CH$_2$)$_{16}$CH$_3$ — Phenyl → CH$_3$CH$_2$CH$_2$COOH — Phenyl

3-Phenyleicosane          Phenylethylacetic acid

(Reprinted with permission from Academic Press, copyright 1962.)


A Pseudomonas putida strain has been shown to oxidize p-xylene to phenolic and catechol compounds (89):

p-xylene → p-xylene dihydrodiol → (nonenzymatic) 2,5-dimethylphenol

→ 3,6-dimethylpyrocatechol

Napthalene compounds undergo similar oxidation sequences as monoaromatics by intermediate dihydrodiol formation on one ring followed by ring fission. Several pathways have been proposed for naphthalene degradation. The following mechanism has been proposed for oxidation by a Pseudomonas sp. (91):

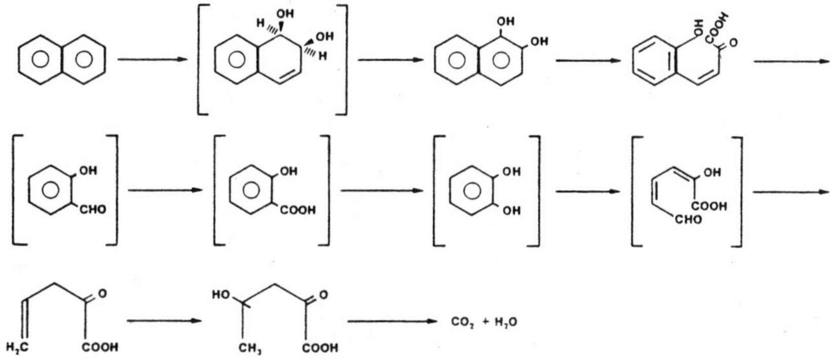

Another sequence for naphthalene oxidation by a Pseudomonas sp. has been proposed by Fernley and Evans (80) (see following page).

One other oxidation sequence for naphthalene has also been proposed (224):

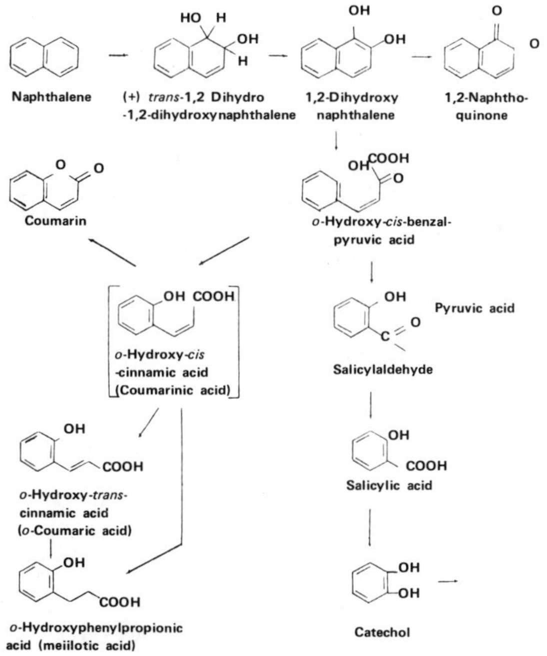

[Reaction scheme showing degradation pathway:]

Naphthalene → D-*trans*-1:2-Dihydronaphthalenediol → 1:2-Dihydroxynaphthalene → 1.2-Naphthaquinone

1:2-Dihydroxynaphthalene → [*o*-Carboxy-*cis*-cinnamic acid] → *o*-Hydroxy-*cis*-cinnamic acid (*o*-Coumarinic acid) ⇌ Coumarin

*o*-Hydroxy-*cis*-cinnamic acid → Salicylic acid → via Catechol

*o*-Hydroxy-*cis*-cinnamic acid → Melilotic acid

*o*-Hydroxy-*trans*-cinnamic acid (*o*-Coumaric acid)

Substituted naphthalenes may be oxidized at the substituent chain or at the aromatic nucleus. Methyl group oxidation results in formation of a methylnaphthoic acid (120), and oxidation of longer chains may produce the corresponding aromatic acid (80):

2-Methylnaphthalene → 2-Naphthoic acid

1-(α-Naphthyl)undecane → 2-(α-Naphthyl)propionic acid

MICROBIAL DEGRADATION    87

Microbial attack of one aromatic ring for alkyl-substituted naphthalenes can result in formation of the corresponding salicyclic acid intermediates (120). Examples for 1- and 2-methylnaphthalene have been provided by a <u>Pseudomonas</u> sp (224):

[Reaction scheme showing:

Row 1: 1-Methyl naphthalene → 8-Methyl-1,2-dihydro-1,2-dihydroxynaphthalene → 3-Methyl-salicylic acid → 3-Methyl-catechol →

Row 2: 2-Methyl-naphthalene → 7-Methyl-1,2-dihydro-1,2-dihydroxynaphthalene → 4-Methyl-salicylic acid → 4-Methyl-catechol → etc.

Branching from 2-Methylnaphthalene: → 2-Naphthylcarbinol → 2-Naphthoic acid

From 7-Methyl-1,2-dihydro-1,2-dihydroxynaphthalene: → 7-Methyl-1-naphthol → 4-Hydroxymethyl-salicylic acid → ?]

Oxidative attack on the unsubstituted ring for these same two compounds has been shown for a <u>Flavobacterium</u> sp. and <u>P. aeruginosa</u>, resulting in catechol formation with ring fission probably occurring between hydroxyl groups (80) (see following page).

Both bromo- and chloro-substituted naphthalenes are degraded by bacteria, with the following pathway proposed for 1-chlornaphthalene by Walker and Wiltshire (80):

1-Methylnaphthalene → [7:8-Dihydro-7:8-dihydroxy-1-methylnaphthalene] → 2-Hydroxy-3-methylbenzoic acid → 3-Methylcatechol

2-Methylnaphthalene → [dihydrodiol intermediate] → 2-Hydroxy-4-methylbenzoic acid → 4-Methylcatechol

(Reprinted with permission of Academic Press, copyright 1962)

MICROBIAL DEGRADATION  89

[Structures: 1-Chloronaphthalene → 1-Chloro-6:7-dihydro-6:7-dihydroxynaphthalene → 3-Chlorosalicylic acid]

(Reprinted with permission of Academic Press, copyright 1962.)

Triaromatic compounds degraded by microbes include anthracene and phenanthrene. Below are the pathways proposed by Rogoff and Wender (80) on degradation by P. aeruginosa:

[Structures: Phenanthrene → 1-Hydroxy-2-naphthoic acid; Anthracene → 3-Hydroxy-2-naphthoic acid; both → Salicylic acid → Catechol]

(Reprinted with permission of Academic Press, copyright 1962.)

A similar but expanded pathway for oxidations of anthracene and phenanthrene has been proposed from several studies utilizing *Pseudomonas* species (224) (see following page).

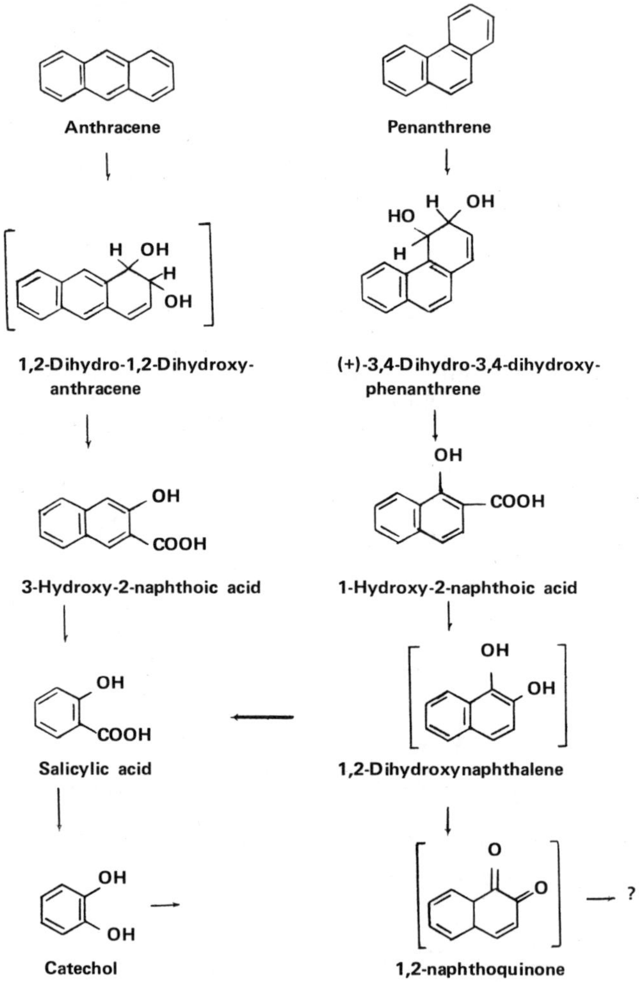

Anthracene ↓

[1,2-Dihydro-1,2-Dihydroxy-anthracene]

↓

3-Hydroxy-2-naphthoic acid

↓

Salicylic acid

↓

Catechol

Penanthrene ↓

(+)-3,4-Dihydro-3,4-dihydroxy-phenanthrene

↓

1-Hydroxy-2-naphthoic acid

↓

[1,2-Dihydroxynaphthalene]

↓

[1,2-naphthoquinone] → ?

(Reprinted with permission from Advances in Enzymology, Vol. 27, copyright 1965 by John Wiley & Sons, Inc.)

A Biejerinekia strain grown on biphenyl substrate has been shown to oxidize biphenyl, anthracene, and phenanthrene to the corresponding dihydrodiols (90):

Biphenyl → *cis* -2,3-dihydroxy-2,3-dihydrobiphenyl

Anthracene → *cis* -1,2,dihydroxy-1,2-dihydroanthracene

Phenanthrene→*cis* -3,4-dihydroxy-3,4-dihydrophenanthrene

This same bacterium also oxidizes some carcinogenic aromatics to vincinal (polar) dihydrodiols (90), such as Benzo($\alpha$)pyrene. The major dihydrodiol of the latter is -9,10-dihydroxy-9,10-dihydrobenzo($\alpha$)pyrene. Benzo($\alpha$) anthracene has four dihydrodiols, with the major one being -1,2-dihydroxy-1,2-dihydrobenzo($\alpha$)anthracene. Other carcinogenic polycyclic aromatics oxidizable by bacteria and found in petroleum at low levels include 1,2-Benzanthracene and 2,2,5,6-Dibenzanthracene (120).

Sulphur aromatics in crude petroleums (thiols, sulfides, and thiophenes) are relatively refractory to microbial metabolism. Thiophenes (condensed) are found in the heavier fractions, and aerobic oxidation of Dibenzothiophene by bacteria has been shown (229) even though the compound is generally toxic to microorganisms. The metabolites were identified as oxygenated derivatives and dihydrodiol intermediates were determined as well (120).

## CONSIDERATIONS FOR USE OF HYDROCARBONOCLASTIC MICROORGANISMS IN OIL SPILL CLEAN-UP

The idea of "seeding" an oil slick with microbial cultures capable of degrading the spill has been proposed, although there is apparent conflict in opinions as to the degree of effectiveness of such procedures. Introduction of "foreign" microbes into a marine environment may pose significant problems in terms of upsetting competitive balances (qualitative and quantitative) within the indigenous microbial population. If a large microbial inoculum is to be introduced into a given environment, the resulting increase in population density will significantly alter localized nutrient availability. It might therefore be necessary to add nutrients concomitantly with the inoculum, which posesses the potential problem of nutrient pollution.

The microbial inoculum could be a mixed culture of bacteria, yeasts, and fungi. Application as an aerosol has been proposed as an effective mode, and may be in the form of a dry power, wet slurry, pelletized, or capsulated with nutrients (49).

## III. PHYSICAL AND ENVIRONMENTAL INTERACTIONS OF SPILLED OIL

### INTERACTIONS WITH SHORELINE ENVIRONMENTS

The fate of spilled oil in coastal environments is primarily controlled by the energy of the environment as defined by exposure to waves and currents and the intertidal substrate receiving the spilled oil residues. In high energy environments (both rocky intertidal zones and sandy beaches) oil is removed quickly, due to wave activity and erosion/accretion of beaches which can alternately expose and cover sedimented oil (27, 45, 100, 104, 125, 181, 189). Even in these high energy environments, however, tar globules of weathered petroleum can be deposited above the splash zone or high tide level where decomposition is relatively slow. Such deposits are sometimes characterized by a crusty outer layer which has been subjected to evaporation and photodecomposition, with an n-alkane distribution within the encrusted ball which is largely unchanged (46).

Thomas (219) demonstrated that following initial oiling after the 1970 Arrow spill in Chedabucto Bay, most stations showed 100% oil cover at the mean high water mark. The oil cover declined at a logarithmic rate, as is shown in Figure III-1 with the losses occurring most rapidly at the lower intertidal levels on shores exposed to the heaviest wave and ice action, and most slowly at higher tidal levels in sheltered locations. The station numbers in the Figure are defined as follows: 1, broken rock and boulder; 2, mainly bed rock with sand at the high water level; 3, broken rock and gravel; 4, bedrock at high water to broken rock at low water; 5, mainly broken rock with sand at the high water level; 6-7, muddy sand at high water to sandy mud at low water. The faster accretion at the lower tidal levels is clear from the figure, and surface oil persisted there only until 1973. Based on the location of the stations in relation to wave action, stations one and two are heavily exposed, stations three and four are moderately exposed, and stations six and seven are sheltered. This figure clearly demonstrates the faster disappearance of oil in proportion to wave exposure. At several of the most sheltered locations, oil still persisted at the mean high water mark as late as 1977.

An example of the differential chemical alterations which occurred in these various zones at Chedabucto Bay is presented in Table III-1. As the data illustrate, the concentrations of true hydrocarbons in the sediments from the low and moderate energy zones remain very similar to a stored sample, whereas the composition in the high energy

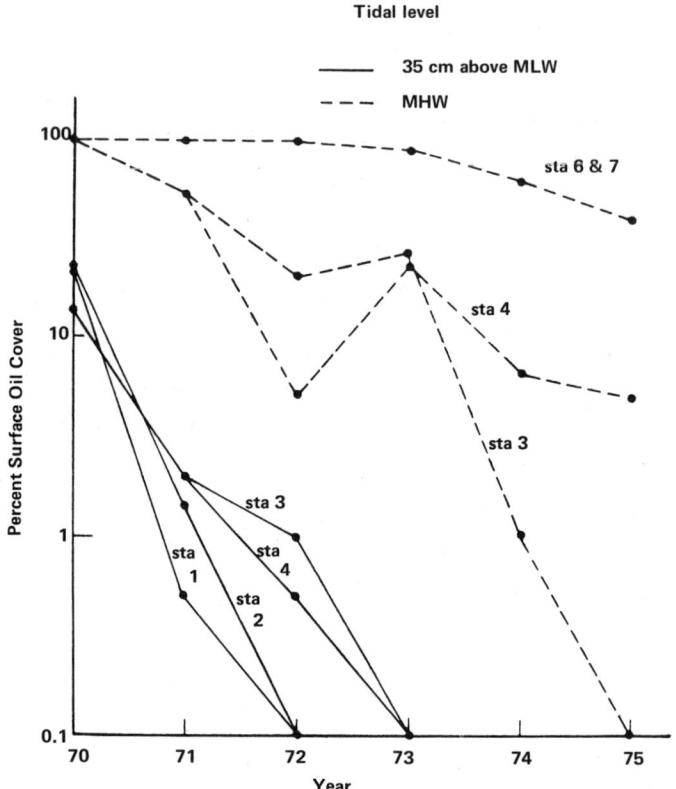

Figure III-1. Surface oil cover at two tidal levels at Chedabucto Bay stations from 1970 to 1975. MHW = mean high water; MLW = mean low water (from Thomas, 219). Reprinted with permission of Pergamon Press Ltd., © 1977).

zone varied considerably, with a significant reduction in aromatic materials and total hydrocarbons (189). The data also show a concomitant increase in non-hydrocarbons (specifically, asphaltenes and resins and nitrogen, sulfur, and oxygen containing materials). The alterations in viscosity and specific gravity as a function of coastal zone energy are also quite dramatic. These viscosity changes clearly reflect the removal of the more volatile and soluble lower molecular weight compounds, leaving only the higher molecular weight residual components. This is particularly evident in the high energy zone as discussed above.

As an example of the evaporative and dissolution losses from light paraffinic crudes in beaches of temperate climates, Blumer et al., (27) showed that after one year, only 10% of the original n-alkanes in the $n-C_{17}$ to $n-C_{18}$ range remained,

Table III-1. Chemical and physical characteristics of Bunker C oil in sediments at Chedabucto Bay. (from Clark and MacLeod, 48; adapted from Rashid, 189). Reprinted with permission of Academic Press, Inc., © 1977.

| Characteristics | Bunker C oil | | Bunker C oil in sediments | | |
|---|---|---|---|---|---|
| | Fresh sample | Stored sample | Low energy coast | Moderate energy coast | High energy coast |
| HYDROCARBONS (%) | | | | | |
| Saturated | [a] | 26 | 25 | 23 | 18 |
| Aromatic | — | 25 | 24 | 24 | 16 |
| Total hydrocarbons | 73.1 | 51 | 49 | 47 | 34 |
| Ratio: saturated/aromatic | — | 1.04 | 1.04 | 0.96 | 1.12 |
| NONHYDROCARBONS (%) | | | | | |
| Asphaltenes | 16.3 | 20 | 22 | 23 | 22 |
| Resins and NSO | 10.6 | 29 | 29 | 30 | 44 |
| Total nonhydrocarbons | 26.9 | 49 | 51 | 53 | 66 |
| Ratio: hydrocarbons/nonhydrocarbons | 2.72 | 1.04 | 0.96 | 0.88 | 0.52 |
| PHYSICAL PROPERTIES | | | | | |
| Specific gravity | 0.950 | 0.963 | 0.9953 | 0.9765 | 0.9823 |
| Viscosity (cP) | — | 19,584 | 28,600 | 1,210,000 | 3,640,000 |

[a] — indicates no data.

with 50% of the $n-C_{19}$ to $n-C_{20}$ components present, and practically all of the $n-C_{23}$ to $n-C_{24}$ hydrocarbons present. In comparison, when oil was spilled in a near-shore or marsh bottom sediment, nearly all of the original hydrocarbons with molecular weights greater than $n-C_{12}$ were found for several years following the spill (25,26,27).

The effect of detergents used for cleaning oiled beaches has been studied (62) following other spill incidents, and it was found in at least one case that direct application of dispersants to oil on beaches was an unsatisfactory method for oil cleanup. In Ubatuba, Brazil (62), an application of BRAS-X-PLUS cosmetically cleaned the beaches, but it also caused the oil to penetrate more deeply into the underlying sand, thus, further distributing it into the interstitial waters. Evidence of detergent-treated oil in beach sand was found seven months later. In many cases the oil penetration on the treated beaches ranged from 7 to 30 cm, and on one beach the penetration was 60 cm. The penetration was primarily a function of the particle size distribution at the treated area; however, the penetration in untreated beaches was limited to 5 cm.

Dispersants work by reducing the oil-water interfacial tension, and thus, when the oil could be broken up more easily, it was transported by water turbulence deeper into the sediments. In addition to expanding the zone of contamination, this greatly complicated the subsequent attempts at mechanical removal and it caused the oil to persist longer in the environment.

Following the Amoco Cadiz oil spill in March of 1978, Hayes et al. (104), completed beach erosion and oil burial studies at 19 sites at the time of the spill and one month later. They determined that the geomorphology in the coastal zone was very significant in the distribution of oil. In many areas it settled in pools around boulders, in bar troughs, in marsh pools, and covered many intertidal rocks, settling in joints and crevices. As in other spills, the exposed rocky coasts were cleaned of relatively heavy doses in a few days of heavy surf. Sheltered rocky coasts, tidal flats and estuarine marsh systems were extremely vulnerable to oil damage, however, and the cleaning was much slower.

Due to the heavy surf following the spill incident (1-2 meter waves were present during the first few days of the spill) and extremely strong tidal currents and spring tides, considerable amounts of oil were buried under the neap berm (up to 25 cm deep) at several of the exposed sandy beaches with time.

A month after the spill much of the oil on the sandy beaches was present as thin bands of small mousse-balls which had been moved by current and wind. In general, however, it was found that the sandy beaches exposed to heavy surf action were cleansed faster than sheltered rocky areas. Coarse cobble beaches changed less, due to the more limited reworking of the intertidal substrate. Nevertheless, it was found that under similar conditions of wave exposure, the sandy beaches were cleaned faster than rocky areas. Considerable evidence was also presented showing high levels of Amoco Cadiz oil in the interstitial waters of the sedimentary regime (104); however, in this study, no chemical characterization of this oil spill was attempted. It was noted that after the beaches were apparently cleaned, there was still considerable oil buried under the berms and associated with the interstitial waters. The digging of pits or entrapment basins during cleanup efforts only further compounded the interstitial water contamination problem.

Because the area affected by the Amoco Cadiz spill had a wide variety of sediment headland and marsh types along with an intricate topography of sheltered coves and exposed areas, the authors generated an oil vulnerability index, which is presented in Table III-2. In that much of the Brittany coast is similar to parts of southern Alaska (104), it may also be desirable to develop a similar approach for that area.

In another study completed after the Amoco Cadiz oil spill, some chemical characterization of mousse and selected environmental samples was attempted (181). In this case, the residual oil on tidal flats was found to be weathered faster than on rock cliffs or at the water line. Aromatics were found to be incorporated into tissues of animals and some oxidation (possibly photo-chemical) of alkylated-dibenzothiophenes to their sulphoxide derivates was detected. Other oxygenated heterocyclic aromatic compounds were not found, however, suggesting a rather short weathering time for the samples analyzed.

In a more extensive set of analyses following the Amoco Cadiz incident, Calder (41) recorded dramatic evidence of significant and different weathering processes on oil deposited in an exposed tidal flat. Gas chromatographic and GC/MS analyses of the hydrocarbons extracted from reference mousse samples and the top centimeter of several sediment samples showed that weathering due to evaporation occurred rapidly, even before the oil was deposited. Ratios of $n-C_{17}$ to $n-C_{18}$ decreased showing evidence of extensive evaporation; however, their losses were greater than those of pristane and phytane, indicating that microbial degradation was also playing a significant role.

Table III-2. An oil spill vulnerability index with particular reference to the Amoco Cadiz oil spill. Higher index values indicate greater long-term damage by the spill (from Mayes et al., 105). Reprinted with permission of the author and the Oil Spill Conference Office, Washington, DC.

| Vulnerability Index | Shoreline Type; Example | Comments |
|---|---|---|
| 1 | Exposed rocky headlands; Douarnenez to Pte. du Raz and Primel-Trégastel to Locquirec. | Wave reflection kept most of the oil offshore; no cleanup needed. |
| 2 | Eroding wave-cut platforms; south of Portsall and F-1 to F-82. | Exposed to high wave energy; initial oiling removed within 10 days. |
| 3 | Fine-grained sand beaches; stations south of Roscoff (AMC-9 & 10) and east of Portsall (AMC-5). | All only lightly oil-covered after 1 month, mainly by new oil swashes. |
| 4 | Coarse-grained sand beaches; AMC-stations 4 (near Portsall) and 12 (St. Cava) and F-38. | Oil coverage and burial after 1 month remains at moderate levels. |
| 5 | Exposed, compacted tidal flats; La Gréve de St. Michel. | No oil remained on the sand flat but did cause enormous mortality of urchins and bivalves. |
| 6 | Mixed sand and gravel beaches; no really good example of this beach type. | The index value is due to rapid oil burial and penetration; all areas had compacted subsurface which inhibited both actions. |
| 7 | Gravel beaches; stations F-80, 95 and 129, also AMC-16. | Oil penetrated deeply (30 cm) into the sediment; cleanup by use of tractors to push gravel into surf zones seemed effective and not damaging to the beach. |
| 8 | Sheltered rocky coasts; common throughout the study area. | Thich pools of oil accumulated in these areas of reduced wave action; cleanup by hand and high-pressure hoses removed some of the oil (this process is valid in nonbiologically active areas). |
| 9 | Sheltered tidal flats; behind Ile Grande and at Castel Meur. | Tidal flats were heavily oiled; cleanup activities removed major oil accumulations, but left remaining oil deeply churned into the sediment; biological recovery yet to be determined. |
| 10 | Salt marshes; Ile Grande marsh. | Extremely heavily oiled with up to 15 cm of pooled oil on the marsh surface; cleanup activities removed the thick oil accumulations but also trampled much of the area; biological recovery yet to be determined. |

Over the seven-month period following the spill, the aliphatic hydrocarbon concentrations decreased in the sediments in an exponential decay fashion, as shown by the data in Figure III-2. During this seven-month period the unresolved complex mixture decreased from 410 ppm to 80 ppm and the n-alkanes in the $n-C_{10}$ to $n-C_{34}$ range decreased from 35 ppm to 1.5 ppm. During the first month the pristane plus phytane concentration dropped from 6.5 to 3 ppm (factor of 2 decrease) and the $n-C_{17}$ plus $n-C_{18}$ concentrations dropped from 1.8 to 0.3 ppm (factor of 6 decrease). The greater loss of the aliphatics compared to the branched isoprenoid compounds clearly demonstrates the importance of the bacterial activity during the first month.

Figure III-3 shows the combined effects of bacterial weathering and evaporation over the seven-month period. The top chromatogram in the figure is the sediment extract collected approximately two weeks after the spill occurred and the bottom chromatogram is the extent obtained after

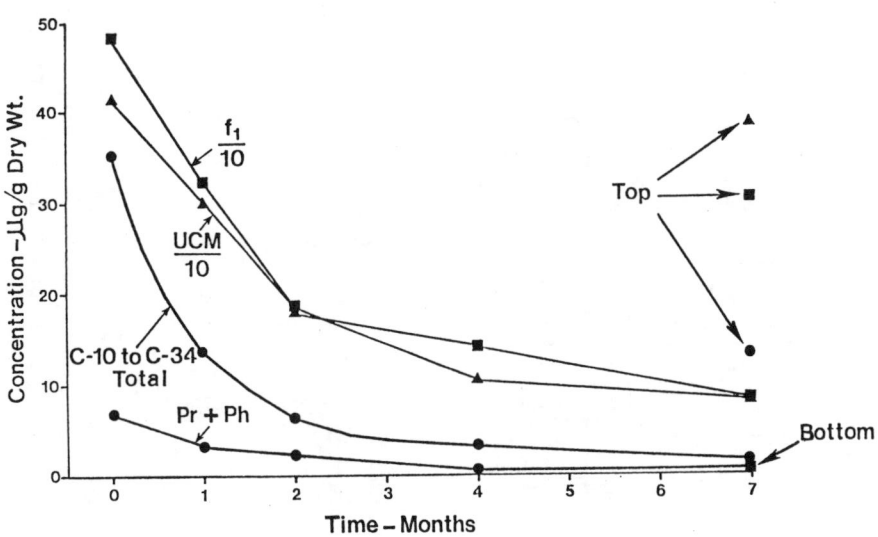

Figure III-2. Decrease in aliphatic hydrocarbon concentrations over time after the <u>Amoco Cadiz</u> oil spill--l'Aber Wrac'h sediment samples from station A (from Calder, 41). Reprinted with permission of the author and the NOAA/OCSEAP Office, Boulder, CO.

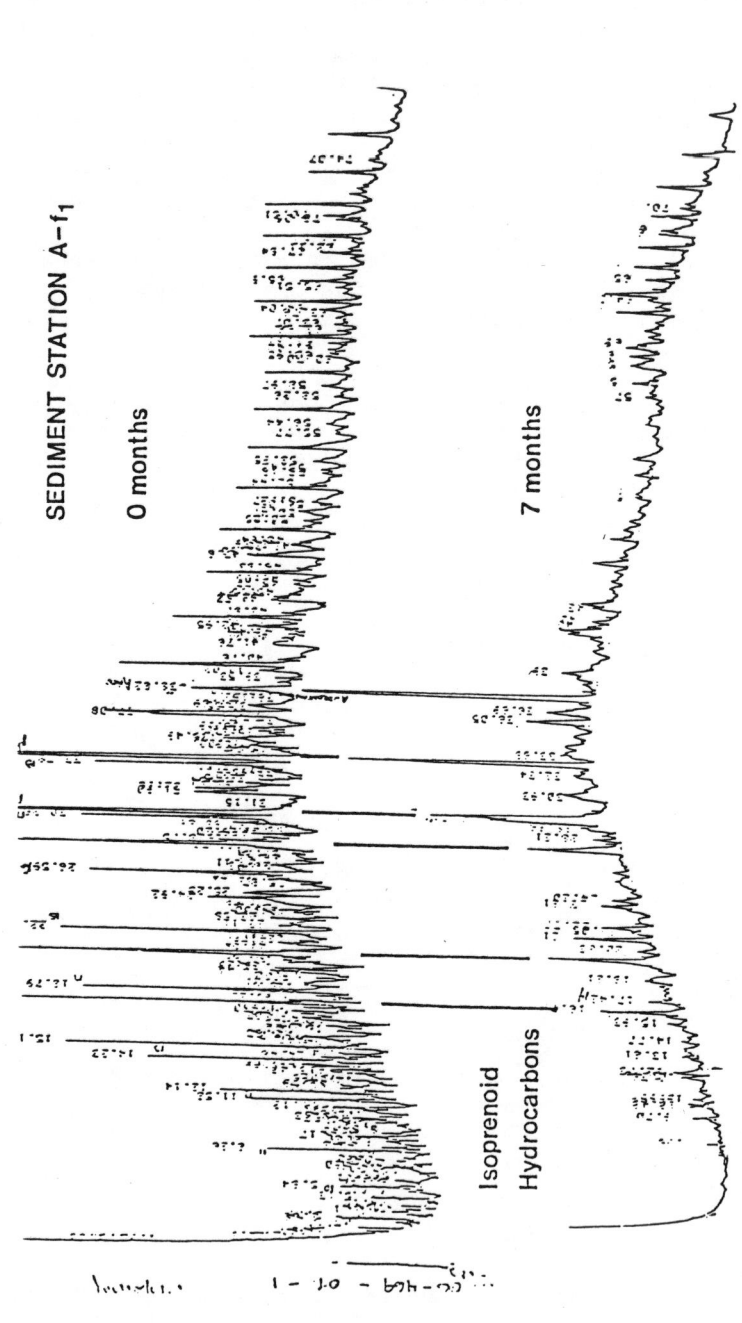

Figure III-3. Glass capillary FID gas chromatograms of the aliphatic fractions of hydrocarbons extracted from sediment samples immediately after the Amoco Cadiz spill and after seven months of weathering (from Calder, 41). Reprinted with permission of the author and the NOAA/OCSEAP Office, Boulder, CO.

seven months. Clearly, all of the n-alkane components are drastically reduced and the isoprenoid hydrocarbons can be seen predominating over the other resolved components in the seven-month old sample. The aromatic hydrocarbons found in the sediments behaved similarly to the aliphatics, and Figure III-4 presents the decrease in concentration of the unsaturate, or aromatic fraction with time as measured by gas-chromatographic analyses. The sum of the resolved components decreased from 29 ppm to 2 ppm over the seven month period. By GC/MS several polynuclear aromatic compounds were detected, including naphthalene, fluorene, phenanthrene and dibenzothiophene. In the samples collected in the early part of April (two weeks after the spill) there was considerable evidence of weathering as naphthalene and fluorene concentrations had dropped considerably from those in the reference mousse. By comparison, the phenanthrene concentrations were approximately 2-3 times greater than the naphthalene concentrations, and the dibenzothiophenes made up well over half of the total resolved components in the sample. In the reference mousse the naphthalenes were by far the most abundant compounds, followed by the fluorene

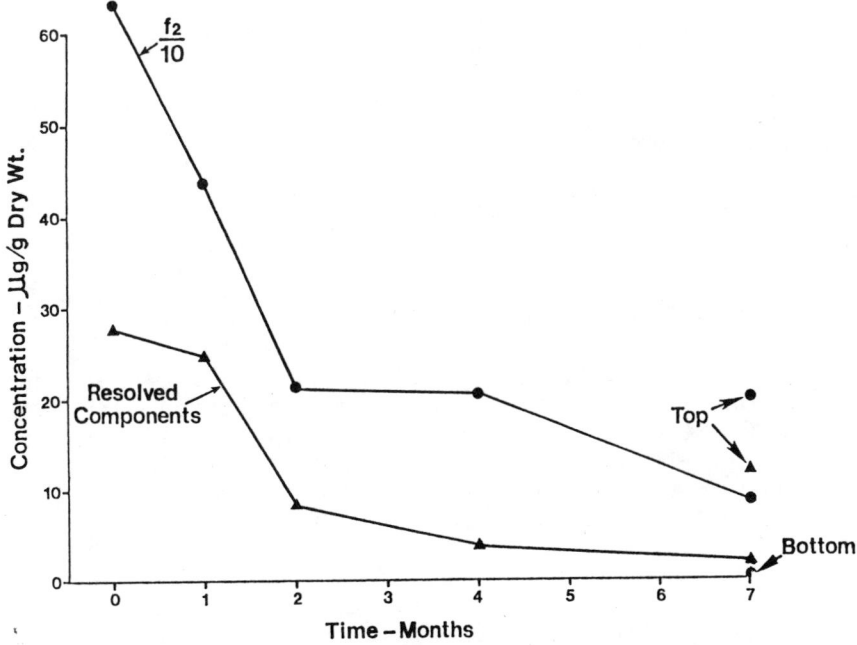

Figure III-4. Decrease in aromatic hydrocarbon concentrations over time after the Amoco Cadiz oil spill--l'Aber Wrac'h sediment samples from Station A (from Calder, 41). Reprinted with permission of the author and the NOAA/OCSEAP Office, Boulder, CO.

and phenanthrene, with dibenzothiophene occurring as "detectable" peaks in the chromatogram. Figure III-5 presents the glass capillary gas chromatograms obtained on the aromatic fractions of the reference mousse and the extent of the upper centimeter of the sediment collected after seven months. Clearly, all of the lower molecular weight aromatics are missing from the seven month old sample with dibenzothiophene remaining as the predominant resolved peak. Concentration changes for the selected aromatics are presented in Table III-3.

Table III-3. Concentrations (μg/g dry wt) of aromatic hydrocarbons in l'Aber Wrac'h, determined by GC/MS ERCO (from Calder, 41). Reprinted with permission of the author and the NOAA/OCSEAP Office, Boulder, CO.

| Compound | Station A at | | Station B at | |
|---|---|---|---|---|
| | 0 mo | 7 mo | 0 mo | 7 mo |
| Naphthalenes | 2.7 | 0.2 | 17.9 | 0.3 |
| Fluorenes | 3.0 | 0.5 | 11.6 | 0.6 |
| Phenanthrenes | 6.7 | 3.2 | 35.8 | 2.8 |
| Dibenzothiophenes | 16.6 | 5.8 | 71.2 | 6.4 |
| Total | 29.0 | 9.7 | 136.5 | 10.1 |

## OIL IN NEAR SHORE SEDIMENTS

Following the Amoco Cadiz spill, oil in the near-shore environment was found to lift off the bottom with every flood tide (for the first few days), only to be redeposited during the ebb tide (104). After one month it remained on the bottom as a water in oil-mousse mixed with sand, and only a light sheen was visible on the water surface during flood tides.

Considerable amounts of oil were detected in the sediments in offshore bottom sediments following the Amoco Cadiz spill (70) and, in general, the distribution showed higher concentrations in fine-grained or muddy sediments and particularly in sediments rich in lithothamnium. In general, the depth

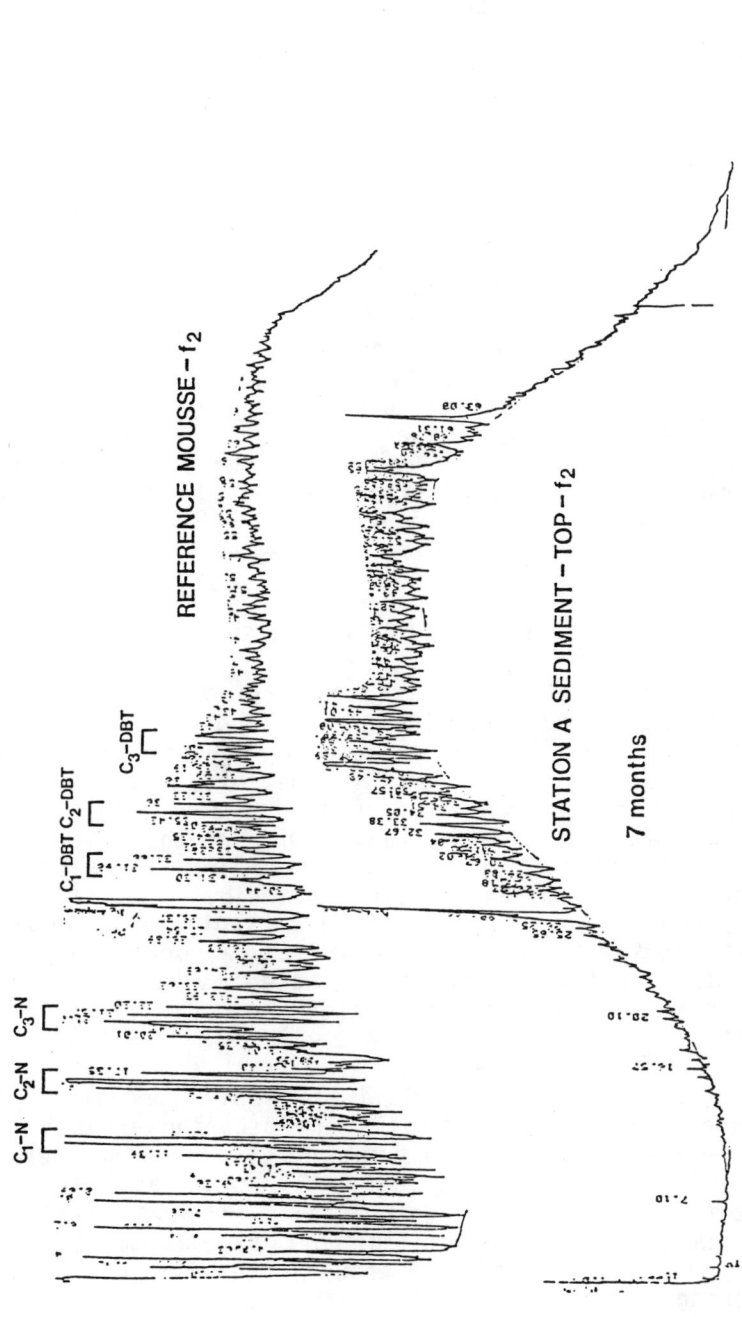

Figure III-5. Glass capillary FID gas chromatograms of the aromatic fraction of hydrocarbons extracted from sediment samples immediately after the Amoco Cadiz spill and after seven months of weathering (from Calder, 41). Reprinted with permission of the author and the NOAA/OCSEAP Office, Boulder, CO.

of penetration was less than 7 cm, possibly due to the limit of biological reworking. Hydrocarbon samples taken in the water column indicated somewhat higher concentrations near the bottom, however, no mechanism of transport information was provided (70). That bottom water current transport is important, has been demonstrated elsewhere (50). In another study on the bottom water current transport of sunken oil in San Francisco Bay (50), the estuarine circulation patterns apparently moved sedimented oil from offshore areas into the back bay estuarine environments.

In tide pool areas it has been demonstrated that light oiling does not affect oxygen transport through the air/oil/water interface. However, the oil may selectively screen ultraviolet or visible light, causing oxygen depletion due to decreased photosynthetic activity. Also, the oil covering in tidepool areas can absorb heat, thus further lowering the oxygen solubility (173). In terms of the effects of oil, these changes are probably only minor compared to its direct toxicity and coating of any organisms involved.

## OIL IN ESTUARINE AND MODERATELY EXPOSED ENVIRONMENTS

In low energy environments, (i.e. protected areas), the sediments tend to be more fine grained and seasonal erosion and accretion patterns are relatively reduced. The slower deposition rate of the low energy sediments allows a longer time for oil to be exposed, and thus before burial, more degradation may occur from evaporation, photooxidation and bacterial activity. Unfortunately, once the oil is sedimented or buried, degradation is inhibited due to lower oxygen concentration (from limited current mixing and the lack of bioturbation if a significant "organism kill" occurred) limiting biological activity to anaerobic organisms at sediment depths greater than a few centimeters (84, 171).

Thus, in contrast to the relatively rapid removal of oil from exposed beaches, oil spills in estuarine areas have a much more drastic impact, and the effects are considerably longer lived (23, 41, 50, 85, 218). In addition, the current regime of a fresh water/saline estuary circulation system can cause entrapment of oil by bottom currents, further depositing oil in the estuary over time, as opposed to cleaning it (50).

In an attempt to approximate a temperate estuary and develop a budget for the rate and transport mechanisms for spilled petroleum, Gearing *et al*. (85) studied the deposition of accommodated Number 2 fuel oil added to tanks at the Marine Ecosystems Research Laboratory (MERL) of the University of Rhode Island. Their results indicated that the main loss of hydrocarbons occurred to the atmosphere with approximately 15% of the oil being adsorbed to the particulate matter and eventually ending in the sediment. Sediment hydrocarbons were depleted in the lower molecular weight aromatics (up to 3 rings), compared to the starting material, again indicating loss due to volatilization. Rates of degradation in the field were developed by comparing GC ratios of n-alkanes to isoprenoids to examine bacterial oxidation, and in laboratory studies by $^{14}C$-labeled hydrocarbons used for monitoring conversion to carbon dioxide. Of the total hydrocarbons from the Number 2 fuel oil, 60-90% were lost due to evaporation, and an exponential decrease of the hydrocarbons in the water fraction was noted (K = 0.0066 hours$^{-1}$). The predominance of the evaporation loss was demonstrated by a constant isoprenoid to n-alkane ratio over 90 days, with an aromatic to n-alkane ratio increasing over that time. This latter phenomenon could be due to preferential evaporation of the lower molecular weight n-alkanes or possibly due to bacterial oxidation. In one tank test, (after 108 days of incubation), additional oil was added. After an additional 90 hours, the n-$C_{17}$/pristane ratio decreased, and the methylnapthalene/(n-$C_{18}$ + n-$C_{19}$) ratio increased, indicating induced microbiological activity. In one experiment completed in this program it was interesting to note that very few aromatic hydrocarbons were detected in the sediments, even after 69 and 131 days, despite detection of aliphatic hydrocarbons. In another tank, however, after 196 and 223 days, some increase in the aromatic hydrocarbon concentration was noted. The aromatic/aliphatic ratio in sediments was much lower than in the parent oil, and the authors concluded that it suggested different mechanisms and/or rates of aliphatic and aromatic deposition throughout these marine ecosystems.

Contradicting results have been reported by other investigators examining hydrocarbon burdens in estuarine or Bay systems (23, 41, 50, 218). Bieri et al. (23) found hydrocarbons in both unconsolidated sediments and in oysters exposed to experimental oil spills. Using wall coated open tubular capillary column GC and GC/MS, the authors noted composition changes of the hydrocarbons with time in both the sediments and tissue matrices. It is interesting to note that in this particular study, chlorinated hydrocarbons were also detected in approximately the same concentrations as the polynuclear aromatic hydrocarbons. Using GC and GC/MS techniques, Teal, Burns and Farrington (218) detected significant amounts of aromatic hydrocarbons in intertidal sediments, from two spills of Number 2 fuel oil in the estuarine environments of Buzzards Bay, Massachusetts. They noted that, as a function of time, the lighter molecular weight naphthalenes were removed first, followed by $C_2$ naphthalenes and $C_3$-naphthalenes. Methylnaphthalenes were the dominant compound in the original oils, however, over a two year period, $C_2$ naphthalenes became dominant in the sediments, followed by the $C_3$ naphthalenes a year later and $C_4$ naphthalenes six years after the spills. In several core samples these lower molecular weight aromatics were detected 10-15 cm below the sediment water interface, even up to four years after the spill. The lighter materials were somewhat degraded at that time, but $C_3$ and $C_4$ naphthalenes were still present.

On the second spill incident, samples were collected three days after the spill, and continued analyses were undertaken for 2 1/2 years. The data of Teal, Burns and Farrington were normalized to the $C_3$ naphthalene concentrations, and they showed that during the first six months all of the groups of aromatics below this range decreased approximately at the same rate. From May 1975 to June 1977, however, the compounds had different rates of loss with the lighter molecular weight materials decreasing faster. They concluded that the heavier molecular weight aromatics were concentrated in the sediments and that they have a longer residence time than the lower molecular weight components. The authors also stated that there was no evidence to indicate that an oil spill from a tanker could convert to the type of compounds expected from combustion sources.

Critney et al. (55) examined sediments six times during a four year period following a spill of heavy fuel oil in a British Columbia coastal bay. Biodegradation accounted for almost complete removal of the n-alkanes in the first year. Branched compounds, pristane and phytane, were degraded more slowly; however, they were completely gone in four years. The non n-alkane components in the n-$C_{28}$ to n-$C_{36}$ range, including several pentacyclic triterpenes were the most resistant to degradation.

This is not the case in anoxic sediments, and Mayo et al. (149) observed that the oxidation of weathering of incorporated petroleum hydrocarbons was much slower in oxygen poor environments. In the cold oxygen-depleted sediments of Long Cove near Searsport, Maine, these authors reported little or no decline in overall hydrocarbon concentrations and practically no weathering of the aliphatic n-alkanes. Their conclusions were that greater weathering occurred in the spill before being transported to the sediments than after it had been buried. In this case, the type of petroleum spilled was a mixture of JP4-jet fuel and Number 2 heating oil. In examinations completed from 1971 to 1976, the compound distributions were virtually identical. Likewise, the absolute amounts of hydrocarbons observed in 1971 and 1976 in the sediments were very similar with average values of 125 ppm and 99 ppm, respectively, for the two dates. Further, the pristane/phytane and $n-C_{17}/n-C_{18}$ concentrations at many of the stations were identical to those obtained in 1971. While these authors did find some evidence of mibrobial degradation of the aliphatic linear alkanes in some of the upland sediments, they did not detect as much microbial degradation in the colder anoxic silt of the cove. One possible explanation for their results could be the additional leaching of unweathered hydrocarbons from the upstream spill site into the anoxic basin; however, the authors did not feel that this possibility explained their results.

Meyers (161) has reported on the distributions of petroleum and indigenous hydrocarbons in recent sediments in Lake Huron. From these results he concluded that virtually no change over a period of several centuries occurs in hydrocarbons buried below a depth of 10 cm. He concluded, therefore, that burial in sediments was an effective sink for petroleum hydrocarbons and sedimentary processes did actively remove them from further interaction with the environment. In his study, however, the presence of large macroinfauna did not allow a full turnover of the upper 10 cm of sediment due to bioturbation, as may be expected in the open ocean. In this situation the fauna were much smaller, and although bioturbation did occur, the reworking of the sediments was limited to a depth of about 5 cm.

In conclusion, several authors have suggested that the sediments are an ultimate sink for petroleum hydrocarbons released into the marine environment (149, 161). Once buried in the sediments, however, the fate depends to a large extent on the environmental conditions at the sediment water interface. In well-oxygenated sediments the presence of various filter feeders, grazers and deposit feeders of the meiofauna and macrofauna leads to the utilization of the organic matter in the sediments. In this process these organisms may expose deeper sediments and organics to the water sediment interface where additional microbial activity can occur.

Clearly, the difference in sedimentary environment is critical to the ultimate degradation and fate of petroleum in spill situations, and as such, the degree of turbulence and oxygen concentration in the sediments and overlying waters is critical for the ultimate fate of the buried oil. The conflicting results of Mayo et al. (149) and those of Calder (41), Cretney et al. (55), Teal et al. (218), clearly illustrate the importance of the chemical and physical parameters associated with the sedimentary regime.

OIL RELEASED IN ARCTIC ENVIRONMENTS; OIL AND ICE/SNOW INTERACTIONS

In general, oil released in Arctic and colder sub-Arctic environments undergoes weathering at much slower rates than oil released in more temperature climates. This is partly due to the low temperatures encountered in the Arctic, and due to the lower ultraviolet light exposure at the higher latitudes. Further, the physical properties of the spilled oil become more important. For example, the pour point of any oil released in a cold environment may be significant.

Prudhoe Bay crude oil has a low enough viscosity to flow over water in summer, but it will solidify in winter (96). When Bunker C fuel oil was released in cold water at 0-2$°$C after the Arrow spill, the oil surfaced in "discrete pieces like a rope, one to three feet long", as it was released through holes in the hull of the vessel (14).

The effects of temperature on the spreading of Prudhoe Bay crude oil on ice has been examined by Fay (77) and the stop-spreading point was determined. Glaeser (95) determined that Prudhoe Bay crude oil spread on zero-degree water to a thickness of 5 mm, but beyond that point, further spreading occurred only by wind. McMinn and Golden (158) determined that some other oils never reach a point where surface tension becomes an important factor in the spreading in cold environments.

In addition to changes in the spreading of oil in cold environments on ice or water, losses due to volatilization are also affected. During the summer in Alaska, Prudhoe crude oil can lose volatile components (up to those boiling below 100$°$C) in 2-5 days with a concomitant increase in viscosity and specific gravity (95). Similar losses occur in winter (15), although at a much slower rate.

Also, as noted above, photochemical reactions in the Arctic do not provide a major removal pathway. This is due to lower ambient light (particularly in winter) and to low oil surface area of exposure due to ice or snow cover. During summer months, photochemical degradation rates may be higher, although this may be attenuated by the reduced intensities due to the angle of incident radiation.

The dispersion of oil when released in cold environments is directly dependent upon the point of release. Oil spilled on top of ice can be trapped in rough, upper ice surface, and in one study about 10%-25% of spilled oil on ice was trapped or absorbed (96). To a certain extent the penetration of the oil depends on the salinity of the ice (19), and thus, on its concomitant porosity. In the Bouchard #65 oil spill in January, 1977, in Buzzard's Bay, the ice was predominantly non-porous and oil penetration was limited to approximately 5-7 cm.

Once on the surface of the ice, the oil can then have a concomitant effect on the ice itself. Oil spilled onto ice absorbs approximately 30% more light than clear ice (95), therefore, this oil can cause the ice to melt during spring at rates of up to 2 cm per day faster than un-oiled ice.

If oil is released beneath ice, it has been demonstrated that most north slope Alaskan crude oils have viscosities low enough to allow flowing easily at the ice/water interface (which is generally at zero degrees regardless of the season). Furthermore, the density of sea ice is approximately 0.85 to 0.91 $g/cm^3$, whereas the underlying water has a density of 1.03 $g/cm^3$. Most Arctic oils have an intermediate density, so they would tend to flow freely under or around the ice, depending upon local conditions.

Thus a different set of entrapment/dispersion scenarios are possible for oil under ice, and these, in large part, depend on the age of the ice. In thick, shorefast ice, depressions under the surface can be +/- 20% of the average ice thickness (176). Thus, oil or gas bubbles once released, can be entrapped in these caverns (138); and once trapped, oil under ice does not weather. Some data indicate that oil has identical viscosity, density, GC profile, etc., after storage under ice (176).

In multi-year ice, pressure ridges as great as 50 meters can occur. These ridges, extending down into the water column, can inhibit flow, and it is conceivable that oil trapped in such areas may remain for up to three or four years (146, 180). A possible recovery method for such trapped oil could entail drilling through the ice and pumping it out (20).

If oil is released at the beginning of the ice growing season, the ice formation is altered by the presence of oil (147). In this situation, a thick, soupy, flexible layer of ice and oil (grease/ice) forms during the early stages of ice growth. In laboratory studies using diesel oil (at $-20^{\circ}C$) when additional diesel oil was released under the grease/ice, it rose rapidly and spread on the surface. There was little evidence of oil adsorption below the grease/ice surface. The grease/ice then continued to grow until it reached a thickness of about 10 cm where, under the influence of artificial waves, (0.6-1.3 meter wave lengths) it broke up into pancake ice. These pancakes grew from the grease/ice and initially floated on the much thicker and denser grease/ice layer. Due to wave activity, grease/ice was pumped onto the ice surface of the pancakes over time by the combination of convergent-divergent motion. This oil which was pumped onto the surface caused ridges to be formed around the edge of the pancake ice. When cold, Prudhoe Bay crude oil was later released under the pancake ice, as viscous globules; it moved through the grease/ice and appeared between the pancake ice. Eventually, due to the wave activities, it was also accumulated within the rims of the pancake ice.

Oil released during winter under first year ice which has already formed will be trapped in brine channels. Then, during the spring and summer meltdown, it can migrate to the surface where it can spread out into the snow over the ice. Oil in snow reduces the albedo thus causing faster meltdown during spring. These oiled pools then continue to melt down into the original lens of pooled oil trapped at the bottom of the brine channels. Once this process has occurred, the ice can then rapidly deteriorate after melt holes develop and surface drainage patterns are established. Because brine channels do not exist in multi-year ice, this penomena is not anticipated to be a factor in oil transport in permanent ice conditions.

It may be possible to further transport oil by a phenomenon known as lead-matrix pumping, which is the opening and closing of leads in the ice (43, 146). Also, oiled hummocks from the closing of leads cause differential melting and oil can thus be further transported in this manner. If pressure ridges are formed with oil-filled or oil-contaminated ice, the movement of the ridge can cause abrasion of the ridge keel, thus releasing oil into non-oiled areas many kilometers from the original site. In multiyear ice, oil might move hundreds or thousands of kilometers in the one to four years that it takes for the oil trapped below the surface to reach the surface of the packed ice.

Oil released under ice during early ice growth has been shown to penetrate 5-10 cm into the loose skeletal layer, where it was then Covered over by additional ice growing below it (176). During the spring melt, it then migrated upwards.

As described above, in laboratory studies, when hot oil was released under ice, the oil broke into globules up to 1.2 mm in diameter and then coalesced under the ice-forming lenses. Melting of the ice did not occur and with continued ice growth the ice surrounded the oil and grew, trapping the oil in pockets (239, 240). Once oil is entrapped by growing ice, cleanup is virtually impossible.

If oil is released in an icepack at the ice/water surface, for example, as a result of a tanker or barge accident, then the oil often flows back along the opening in the ice formed by the barge and tug (19). This situation is discussed later in this section under a detailed consideration of the Bouchard #65 oil spill in Buzzards Bay in January, 1977.

If oil is spilled into snow, a variety of physical interactions occur. The oil can fill the void volume in snow up to 20% due to gravity, followed by capillary action which further spreads the oil (139). If hot oil is spilled onto snow, this first causes melting with an oil/water mixture penetrating further where the water eventually refreezes, halting the downward flow of the oil in pools (139, 207). Additional snow coverage by blowing wind and snow drift or increased snowfall then drastically reduces weathering of the oil by evaporation (158). Light energies can penetrate the snow, however, and thus oil under snow can absorb energy at higher rates than non-oiled snow (176). Following the Buzzard's Bay oil spill in January of 1977, a light snowfall completely covered the oiled areas, but within two days the oil had formed a mulch with the snow and resurfaced (19).

The Bouchard #65 oil spill in Buzzard's Bay in January 1977 provided scientists with a unique opportunity to study a moderate sized spill in an ice covered environment under natural conditions (19). For that reason, the spill has received somewhat greater attention (compared to several others) and it will be considered in somewhat greater detail in the following section.

After the grounding of the barge Bouchard #65 on Cleveland Ledge in Buzzard's Bay, Massachusetts on 28 January, 1977, its cargo of Number 2 heating oil immediately began spilling into the bay. Eventually, a total of 81,146 gallons of heating oil was spilled, and much of this oil filled the channels left by the barge and tug and then further accumulated in the brash ice left in the wake of these vessels and the Coast Guard cutter Towline. Approximately 3,600 gallons of Number 2 heating oil were trapped in this manner following this spill. (During subsequent cleanup efforts, approximately 50% of this oil was successfully burned.) Oil also built up against the edges of ice flows to a depth of approximately 5 cm and then flowed beneath the smooth ice to brash ice downstream.

Uzener et al. (223) demonstrated that current velocities of 0.035 meters/second are sufficient to initiate motion of No. 2 diesel fuel under smooth ice. Once this under ice transport has started, the oil can then be moved at a velocity $V_0$ which is defined by $V_0 = [3.8V_c - 0.0133]$ meters/second, where $V_c$ is the current velocity in meters/second.

This relationship apparently holds up to current velocities of 0.33 meters/second. In the Bouchard #65 spill, the smooth surface of the undersea ice prevented formation of oil lenses, as described above, for permanent ice or first year ice. This, no doubt, reflected the age of the ice as well as the conditions of its growth. In the Buzzard's Bay area only two months of ice growing season existed at the time of the Bouchard spill. Therefore, most of the ice was very thin (0.3 meters thick, although ridging and rafting in the center of Buzzard's Bay did increase the thickness of some ice to one meter). The ice in this area was also very fresh (salinity ranged from 2-4 $^0/oo$), so it was not as porous as Arctic ice may be. In numerous sections of the Buzzard's Bay area, current velocities of 2.3 meters/second occur, so much of the oil was entrained under the ice and transported until it rose into openings in brash ice or was trapped in rafted ice where it was protected from the current. After three days the oil was transported a maximum of 4.3 km from the original spill site. Figure III-6 shows an illustration of oil entrapment in the lee side of rafted ice. There, it was protected from the currents and not moved. During cleanup efforts, oil in such pools in rafted ice was removed by pumping.

Because of their virtual absence in the icepack, no oil concentrated against any ridge keels or pools. Some pressure ridges did contain oiled blocks of ice, but oil penetration ranged from only 1-3 cm into the surface. Rubble fields were more porous and additional oil was trapped there. Also, significant amounts of oil were blown onto the surface of pancake ice, where low porosity and strong winds forced a gradual spreading. Penetration was generally limited to less than 3 mm (19). Due to a 75% decrease in the albedo level, however, partial melting of the pancake ice occurred during the day with oil/water pools then refreezing in a fishnet pattern in the evening. Shoreline fast ice served as a cover and oil migrated under the ice to the beach areas, where in some cases the oil saturated the sediments to a depth of 10 cm.

PHYSICAL & ENVIRONMENTAL INTERACTIONS    113

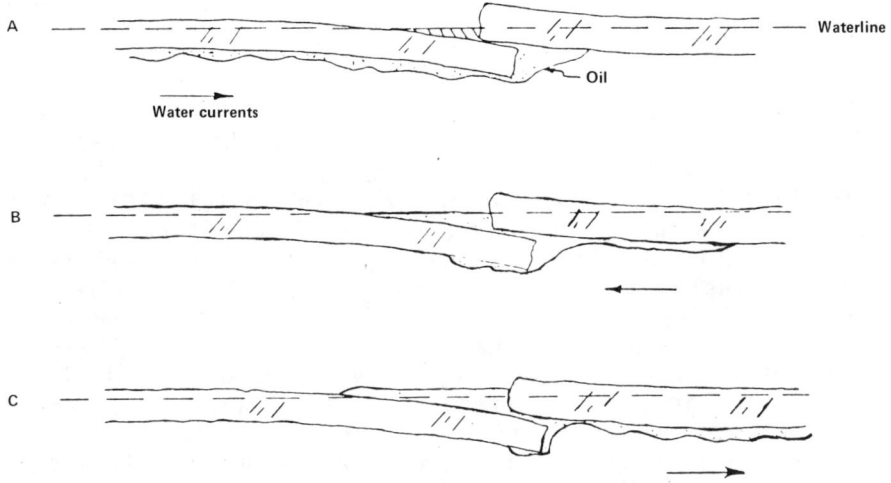

Figure III-6. Flow of oil in rafted ice. (a) Oil flowing under the ice comes in contact with rafted ice; (b) current reversal encourages oil filling into rafted ice pocket; and (c) reversal of current sweeps unsheltered oil away, 8-9 February (from Baxter et al., 19). Reprinted with permission of the NOAA/Science Applications, Inc. MESA Program, Boulder, CO.

During breakup of the ice after the Buzzard's Bay spill, oiled ice was transported out of the region where it melted further at sea. As a result of this activity, the extent of oil spreading in some of the coastal areas appears to have been minimized.

As noted before, the Bouchard #65 spill was extensively studied by a variety of scientists, and as a result, the effects of oil weathering due to evaporation were closely examined. Table I-1 shows that the percent of loss due to evaporation and other weathering effects is clearly a function of field conditions (19). Aromatics appear to be preferentially lost with up to 14-89% of the aromatics being evaporated, and aliphatics only changing from 4% to 37%, depending on the conditions. (It should be remembered, though, that the cargo of the Bouchard #65 consisted of primarily saturated hydrocarbons, ranging from n-$C_9$ to n-$C_{23}$ for 67%, with the remaining 33% consisting of from 1 to 3 ring substituted aromatic compounds). Oil underneath the ice (or near the edge of rafting ice) underwent the least amount of weathering, with oil on top of ice or oil on ice chunks rotated in air, undergoing the greatest alterations. The snowfall which occurred several days after the spill caused the formation of mulch/slush, which contained up to 30% oil by volume, and it was almost impossible to clean up this material.

A good summary of cleanup procedures is presented in reference (19) for the Bouchard spill and also in reference 20, which deals with the interaction of diesel fuel with freshwater ice. Freshwater ice can contain 60% oil by volume, and being much weaker than uncontaminated ice, it will melt first during the spring. Attempts at burning the diesel fuel oil on the freshwater spill met with some success. However, as with the case of the Bouchard spill, considerable difficulty was encountered in getting the oil to light. Heat did cause capillary action on the oil in the freshwater ice, thus perpetuating the initial burn, as the oil was removed from the ice. However, heavy snow or wind prevented wicking or burning of the surface oil in several instances. In the case of the grounding of the Imperial St. Clair (20) and the cleanup of the Bouchard #65 oil spill (20), cleanup methods became almost ineffective as soon as the ice broke into floes and ice movement occurred.

## IV. PROPERTIES AND TYPES OF CRUDE OILS AND PETROLEUM PRODUCTS

Several excellent reviews have been published on the chemical and physical properties of crude oils and refined petroleum hydrocarbon products (47, 48, 156, 237). Rather than attempt to repeat those efforts, we will briefly discuss only the more significant properties which relate to the fate of spilled oil. For more details, the reader is referred to the cited reviews.

In terms of the distribution of oil once it is spilled into the marine environment the pour point is extremely important. By definition, the pour point is the lowest temperature at which an oil can be poured. Dean (60) has listed the pour points of six representative crude oils and values ranged from $-35°C$ to $7°C$. This range can extend from $-43°C$ to $+43°C$ (1) when 93 of the world's major crude oil blends are considered, and Table IV-1 shows the pour points for 9 commercially produced crude oils and gas oils with a range extending from $-42°$ to $+38°C$ (237).

Table IV-1. The pour points for common crude oils and selected gas oil fractions (from Wheeler, 237).

| Oil | Crude (°C) | Gas Oil (°C)[a] |
|---|---|---|
| Abu Dhabi, Murban | -12 | 38 |
| Prudhoe Bay | -9 | |
| Arabian Light | -26 | 32 |
| Arabian Heavy | -34 | 32 |
| Iranian Light | -12 | 35 |
| Iranian Heavy | -21 | 34 |
| Tiajuana Light | -43 | |
| Kuwait | -20 | |
| Ekofisk | -5 | |

[a] $343-565°C$ fraction, average carbon number $C_{28.5}$.

Thus, if petroleum is spilled into the environment at a temperature less than its pour point, the spilled material would tend to become cohesive and this could simplify the cleanup activities. Also, the pour point is of significance when a fuel oil or lubricant is used in machinery which may be exposed to extremely low temperatures.

The boiling point ranges of petroleum are also significant, particularly with regard to losses due to evaporation. Figure IV-1 illustrates the boiling point range of a variety of petroleum products.

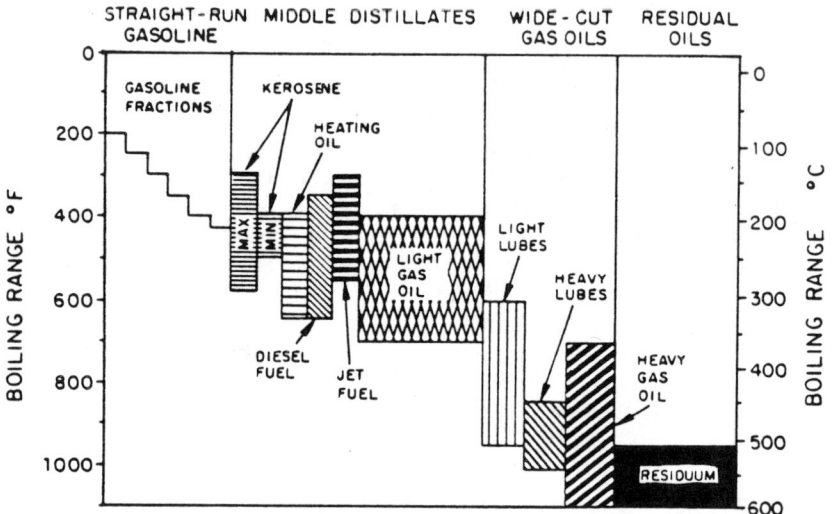

Figure IV-1. Boiling point range of fractions of crude and refined petroleum (from Clark and Brown, 47). Reprinted with permission of Academic Press, Inc., © 1977.

The compositional characteristics of a variety of crude oils and refined products have been studied extensively, and Clark and Brown (47) have reviewed these properties and their importance in spilled oil scenarios. Table IV-2 presents typical analyses of Prudhoe Bay crude oil and fractions which are obtained by refining Prudhoe Bay crude oils. Table IV-3, which was also taken from Clark and Brown (47), presents physical characteristics and chemical properties of 3 different crude oils and allows a comparison of such factors as their specific gravity, trace element composition and the relative proportions of aliphatic, naphthenic and aromatic compounds. Table IV-4 presents different refinery fractions by hydrocarbon types and again illustrates the boiling point range as a function of the number of carbon atoms. The different physical characteristics of refined products are reviewed in Table IV-5, which contrasts the properties of a Number 2 Fuel Oil and a Bunker C fuel oil. Clearly, the variance in the API gravity illustrates the difference in density which is also reflected by: the presence or absence of lower molecular

Table IV-2. Analysis of typical Prudhoe Bay crude oil fractions (from Clark and Brown, 47). Reprinted with permission of Academic Press, Inc., 1977.

| Characteristic or Component | Crude Oil | Natural Gas | Naphtha Gasoline | Naphtha Kerosene | Middle Distillate | Wide-Cut Gas Oils | Residuum |
|---|---|---|---|---|---|---|---|
| Boiling Point Range, °C |  | <20 | 20-190 | 190-205 | 205-343 | 343-565 | 565+ |
| Specific Gravity (15°C) | 0.8883 |  | 0.7531 | 0.818 | 0.8581 | 0.9279 | 1.0231 |
| API Gravity, °API | 27.8 |  | 56.9 | 41.5 | 33.4 | 21.0 | 6.8 |
| Pour Point, °C | -10 |  |  | <-60 | -23 | 35 | 52 |
| Viscosity |  |  |  |  |  |  |  |
| Saybolt (38°C),[a] sec | 73.5 |  |  |  | 36.1 | 85-200 | >200 |
| Kinematic (38°C), cSt | 14.0 |  |  |  | 3.05 | >30 |  |
| Yield, vol% |  |  |  |  |  |  |  |
| Crude Oil | 100 | 3.08 | 18.0 | 2.1 | 24.6 | 35.0 | 17.6 |
| Paraffins | 27.3 | 100 | 47.3 | 41.9 | 8.9 | 9.3 |  |
| Naphthenes | 36.8 | 0 | 36.8 | 38.1 | 14.4 | 22.8 |  |
| Aromatics | 25.3 | 0 | 15.9 | 20.0 | 76.7[b] | 67.9[b] |  |
| Others | 10.6 | 0 | 0 | 0 |  |  |  |
| Composition |  |  |  |  |  |  |  |
| Sulfur, wt% | 0.94 |  | 0.011 | 0.04 | 0.34 | 1.05 | 2.30 |
| Mercaptan Sulfur, ppm | 20 |  | 5 |  |  |  |  |
| Nitrogen, wt% | 0.23 |  | 0.02 | 0.02 | 0.04 | 0.16 | 0.68 |
| Oxygen, wt% | 0.01 |  |  |  |  |  |  |
| Vanadium, ppm | 18 |  | 0 | 0 | 0 | <1 | 93 |
| Nickel, ppm | 10 |  | 0 | 0 | 0 | <1 | 46 |
| Iron, ppm | 4 |  | 0 | 0 | 0 | <1 | 25 |

[a] Saybolt viscosity = the time in seconds for 60 ml of a sample to flow throught a calibrated Universal orifice under specified conditions, according to ASTM method D-88 (10:part 23).
[b] Includes naphtheno-aromatic compounds and nonhydrocarbons.
[c] Polar compounds, nonvolatile aromatic hydrocarbons and column holdup in fractions boiling at 205°C.

Table IV-3. Physical characteristics and chemical properties of several crude oils (from Clark and Brown, 47). Reprinted with permission of Academic Press, Inc., 1977.

| Characteristic or Component | Prudhoe Bay[a] | South Louisiana[b] | Kuwait[b] |
|---|---|---|---|
| API Gravity (20°C), °API | 27.8 | 34.5 | 31.4 |
| Sulfur, wt% | 0.94 | 0.25 | 2.44 |
| Nitrogen, wt% | 0.23 | 0.69 | 0.14 |
| Nickel, ppm | 10 | 2.2 | 7.7 |
| Vanadium, ppm | 20 | 1.9 | 28 |
| Naphtha Fraction, wt%[c] | 23.2 | 18.6 | 22.7 |
| Paraffins | 12.5 | 8.8 | 16.2 |
| Naphthenes | 7.4 | 7.7 | 4.1 |
| Aromatics | 3.2[d] | 2.1 | 2.4 |
| Benzenes | 0.3 | 0.2 | 0.1 |
| Toluene | 0.6 | 0.4 | 0.4 |
| $C_8$ Aromatics | 0.5 | 0.7 | 0.8 |
| $C_9$ Aromatics | 0.06 | 0.5 | 0.6 |
| $C_{10}$ Aromatics | | 0.2 | 0.3 |
| $C_{11}$ Aromatics | | 0.1 | 0.1 |
| Indans | | | 0.1 |
| High-Boiling Fraction, wt%[e] | 76.8[f] | 81.4 | 77.3 |
| Saturates | 14.4[g] | 56.3 | 34.0 |
| n-Paraffins | 5.8 | 5.2 | 4.7 |
| $C_{11}$ | 0.12 | 0.06 | 0.12 |
| $C_{12}$ | 0.25 | 0.24 | 0.28 |
| $C_{13}$ | 0.42 | 0.41 | 0.38 |
| $C_{14}$ | 0.50 | 0.56 | 0.44 |
| $C_{15}$ | 0.44 | 0.54 | 0.43 |
| $C_{16}$ | 0.50 | 0.58 | 0.45 |

## PROPERTIES & TYPES

| | | | |
|---|---:|---:|---:|
| $C_{17}$ | 0.51 | 0.59 | 0.41 |
| $C_{18}$ | 0.47 | 0.40 | 0.35 |
| $C_{19}$ | 0.43 | 0.38 | 0.33 |
| $C_{20}$ | 0.37 | 0.28 | 0.25 |
| $C_{21}$ | 0.32 | 0.20 | 0.20 |
| $C_{22}$ | 0.24 | 0.15 | 0.17 |
| $C_{23}$ | 0.21 | 0.16 | 0.15 |
| $C_{24}$ | 0.20 | 0.13 | 0.12 |
| $C_{25}$ | 0.17 | 0.12 | 0.10 |
| $C_{26}$ | 0.15 | 0.09 | 0.09 |
| $C_{27}$ | 0.10 | 0.06 | 0.06 |
| $C_{28}$ | 0.09 | 0.05 | 0.06 |
| $C_{29}$ | 0.08 | 0.05 | 0.05 |
| $C_{30}$ | 0.08 | 0.04 | 0.07 |
| $C_{31}$ | 0.08 | 0.04 | 0.06 |
| $C_{32}$ plus | 0.07 | 0 | 0.06 |
| iso-Paraffins | | 14.0 | 13.2 |
| 1-ring cycloparaffins | 9.9 | 12.4 | 6.2 |
| 2-ring cycloparaffins | 7.7 | 9.4 | 4.5 |
| 3-ring cycloparaffins | 5.5 | 6.8 | 3.3 |
| 4-ring cycloparaffins | 5.4 | 4.8 | 1.8 |
| 5-ring cycloparaffins | | 3.2 | 0.4 |
| 6-ring cycloparaffins | | 1.1 | |
| Aromatics, wt% | 25.0 | 16.5 | 21.9 |
| Benzenes | 7.0 | 3.9 | 4.8 |
| Indans and tetralins | | 2.4 | 2.2 |
| Dinaphtheno benzenes | | 2.9 | 2.0 |
| Naphthalenes | 9.9 | 1.3 | 0.7 |
| Acenaphthenes | | 1.4 | 0.9 |
| Phenanthrenes | 3.1 | 0.9 | 0.3 |
| Acenaphthalenes | | 2.8 | 1.5 |

Table IV-3, continued

| Characteristic or Component | Prudhoe Bay[a] | South Louisiana[b] | Kuwait[b] |
|---|---|---|---|
| Pyrenes | 1.5 | | |
| Chrysenes | | | 0.2 |
| Benzothiophenes | 1.7 | 0.5 | 5.4 |
| Dibenzothiophenes | 1.3 | 0.4 | 3.3 |
| Indanothiophenes | | | 0.6 |
| Polar materials, wt%[h] | 2.9 | 8.4 | 17.9 |
| Insolubles, wt%[i] | 1.2 | 0.2 | 3.5 |

These analyses represent values for one typical crude oil from each of the geographical regions; variations in composition can be expected for oils produced from different formations or fields within each region.

[a]Adapted from Thompson et al. (23) and Coleman et al. (24)
[b]From Pancirov (25).
[c]Fraction boiling from 20 to 150°C.
[d]Reported for fraction boiling from 20 to 150°C.
[e]Fraction boiling above 205°C.
[f]Reported for fraction boiling above 220°C.
[g]Prudhoe Bay crude oil weathered two weeks to duplicate fractional distillation equivalent to approximately 205°C n-percents from gas chromatography over the range $C_{11}$ to $C_{32}$ plus for the Prudhoe Bay sample only. Unpublished data (R. C. Clark, Jr.).
[h]Polar material: clay-gel separation according to ASTM method D-2007 (10; part 24) using pentane on unweathered sample.
[i]Insolubles: pentane-insoluble materials according to ASTM method D-893 (10; part 23).

Table IV-4. Refinery fractions by hydrocarbon types from crude petroleum (from Clark and Brown, 47). Reprinted with permission of Academic Press, Inc., © 1977.

| Distillation Fraction | Approximate Boiling Range (°C) | Hydrocarbon Types | Range of Carbon Atoms | Typical Refined Products |
|---|---|---|---|---|
| Natural Gas | <20 | Paraffins | 1-6 | Natural Gas |
| Gasoline and Naphtha | 20-200 | Paraffins Aromatics Naphthenes | 4-12 | Gasoline |
| Middle Distillate | 185-345 | Paraffins Aromatics Naphthenes | 10-20 | Kerosene Jet Fuel Heating Oils Diesel Oils |
| Wide-Cut Gas Oil | 345-540 | Paraffins Aromatics Naphthenes | 18-45 | Catalytic Cracking Feedstock Lube Oils Wax |
| Residuum | <540 | Complete Aromatic and Naphthenic Compounds | <40 | Residual Oils Asphalt Coke |

Table IV-5. Physical characteristics and chemical properties of two refined products (from Clark and Brown, 47). Reprinted with permission of Academic Press, Inc., © 1977.

| Characteristic or Component | No. 2 Fuel Oil | Bunker C Fuel Oil |
|---|---|---|
| API Gravity (20°C), °API | 31.6 | 7.3 |
| Sulfur, wt% | 0.32 | 1.46 |
| Nitrogen, wt% | 0.024 | 0.94 |
| Nickel, ppm | 0.5 | 89 |
| Vanadium, ppm | 1.5 | 73 |
| Saturates, wt% | 61.8 | 21.1 |
| n-paraffins | 8.07 | 1.73 |
| $C_{10} + C_{11}$ | 1.26 | 0 |
| $C_{12}$ | 0.84 | 0 |
| $C_{13}$ | 0.96 | 0.07 |
| $C_{14}$ | 1.03 | 0.11 |
| $C_{15}$ | 1.13 | 0.12 |
| $C_{16}$ | 1.05 | 0.14 |
| $C_{17}$ | 0.65 | 0.15 |
| $C_{18}$ | 0.55 | 0.12 |
| $C_{19}$ | 0.33 | 0.14 |
| $C_{20}$ | 0.18 | 0.12 |
| $C_{21}$ | 0.09 | 0.11 |
| $C_{22}$ | 0 | 0.10 |
| $C_{23}$ | 0 | 0.09 |
| $C_{24}$ | 0 | 0.08 |
| $C_{25}$ | 0 | 0.07 |
| $C_{26}$ | 0 | 0.05 |
| $C_{27}$ | 0 | 0.04 |
| $C_{28}$ | 0 | 0.05 |
| $C_{29}$ | 0 | 0.04 |
| $C_{30}$ | 0 | 0.04 |
| $C_{31}$ | 0 | 0.04 |
| $C_{32}$ plus | 0 | 0.05 |
| iso-Paraffins | 22.3 | 5.0 |
| 1-ring cycloparaffins | 17.5 | 3.9 |
| 2-ring cycloparaffins | 9.4 | 3.4 |
| 3-ring cycloparaffins | 4.5 | 2.9 |
| 4-ring cycloparaffins | 0 | 2.7 |
| 5-ring cycloparaffins | 0 | 1.9 |
| 6-ring cycloparaffins | 0 | 0.4 |
| Aromatics, wt% | 38.2 | 34.2 |
| Benzenes | 10.3 | 1.9 |
| Indans and Tetralins | 7.3 | 2.1 |
| Dinaphtheno Benzenes | 4.6 | 2.0 |
| Naphthalene | 0.2 | |
| Methylnaphthalenes | 2.1 | 2.6 |
| Dimethylnaphthalenes | 3.2 | |
| Other Naphthalenes | 0.4 | |
| Acenaphthenes | 3.8 | 3.1 |
| Acenaphthalenes | 5.4 | 7.0 |
| Phenanthrenes | 0 | 11.6 |
| Pyrenes | 0 | 1.7 |
| Chrysenes | 0 | 0 |
| Benzothiophenes | 0.9 | 1.5 |
| Dibenzothiophenes | 0 | 0.7 |
| Polar Materials, wt% | 0 | 30.3 |
| Insolubles (Pentane), wt% | 0 | 14.4 |

These analyses represent typical values for two different refined products; variations in composition can be expected for similar materials from different crude oil stocks and different refineries.

[a] This is a high aromatic material; a typical No. 2 fuel oil would have an aromatic content closer to 20-25%.

weight n-alkane components in the Number 2 fuel oil versus the concentration to 1 to 6-ring cycloparaffins in the Bunker C fuel oil, and the relatively greater abundance of polynuclear aromatic hydrocarbons, including alkyl substituted naphthalenes, phenanthrenes, pyrenes, and sulphur containing heteroaromatic compounds (including benzothiophenes and dibenzothiophenes) in the Bunker C oil. The relative differences in insoluble material in the fuel oil and Bunker C are also illustrated in the Table IV-5.

These compositional characteristics and physical properties are important when oil is released into the environment because the presence of nitrogen, sulfur and oxygen compounds generally increases the water solubility of the petroleum compounds, and along with the viscosity, can be significant in the formation of both oil-in-water and water-in-oil emulsions. The presence of NSO compounds is also important in that some of their degradation products are more toxic than the parent materials. Conversely, high sulfur containing crudes or refined products may not be as rapidly degraded by photochemical oxidation, because of the organo-sulfur compounds ability to act as a free radical chain terminating material via the formation of sulphoxides.

The significance of the specific gravity and changes in specific gravity and viscosity with weathering are important with regard to the dispersion, dissolution and sinking of oil once it is released into the marine environment. Immediately upon its release, weathering processes affect the oil and Table I-4 presents time series density and kinematic viscosity alterations as a function of weathering time for two crude samples. Extremely cold temperatures can also induce viscosity changes, and Table IV-6 shows the viscosity alterations versus time at three different temperatures. Time factors could affect the spreading of oil if released on ice or into cold Arctic waters.

For example, the density of undeformed sea ice depends on its salinity and porosity, but averages about $0.9$ $g/cm^3$. The density of seawater is approximately 1.03 $g/cm^3$. Thus, if oil of higher density than the ice but less dense than seawater is released, it will remain on the water surface until ice freezes over it and it is forced under by waves or currents where further weathering would be inhibited. In ice-free situations where surface weathering can occur, volatile and soluble components are rapidly removed. The resultant residues of most crudes and refined products then typically have greater specific gravities than the starting material. Kuwait crude which has a specific gravity of

Table Iv-6. Viscosity alterations as a function of time and temperature (from Wheeler, 237). Reprinted with permission of the author and the Exxon Production Research Co.

| Elapsed Time (min) | Viscosity (mPa-sec)[a] | | |
|---|---|---|---|
| | $-3°C$ | $-8°C$ | $-14°C$ |
| 0 | 9.6 | 13.0 | 18.1 |
| 10 | 11.4 | 14.2 | 19.8 |
| 20 | 13.8 | 15.8 | 20.9 |
| 30 | 15.0 | 17.9 | 22.0 |
| 40 | 16.8 | 18.9 | 23.7 |
| 50 | 18.7 | 20.0 | 24.5 |

[a] mPa-sec = centipoise.

about 0.87 g/cm$^3$, has a specific gravity of 1.023 after removal of the fractions boiling below 520°C. Iranian heavy crude oil has a specific gravity of 0.869 g/cm$^3$ and its residue after evaporation and weathering has a specific gravity of 1.027. Therefore, residues of these two particular types of oil could sink or tend to disperse readily, either by emulsification or sinking, after weathering has occurred (82). Most Arctic oils are more dense than either sea ice or fresh water ice, and as a result of that, they will tend to flow under ice if released in the Arctic environment.

## V. APPROACHES TO MASS BALANCE PROBLEMS

To fully understand and predict the fate and impact of oil spilled in the marine environment, attempts must be made to calculate an overall oil budget or mass balance following the spill. This analysis aids in the description of the types of oil substrate interactions which leads towards designing workable predictive models for spilled oil transport; it improves the effectiveness of cleanup operations by allowing the estimation of the amount of accessible oil; and it serves as a basis for developing more effective data on collection techniques and improved cleanup operations.

Various mass balance calculations have been attempted using real spill situations and controlled ecosystem or laboratory experiments, as well as computer models simulating oil spills. In this section we briefly review the results of such studies on oil in intertidal (rocky or sandy beach) environments, the open ocean, estuarine systems, and ice covered regions (19, 54, 85, 104).

Following the wreck and dispersion of oil from the Amoco Cadiz, (March 16 - March 30, 1978), field studies were undertaken to attempt a mass balance estimate of the total oiled beach on March 19 - April 2 and again during the period of April 20 - April 28, 1978 (104). This field work consisted of overflights and intensive ground inspection and surveys of the entire affected area at 19 permanent beach survey stations and 147 observation stations (104). During the first two weeks of the spill a total of 72 kilometers of coast was heavily oiled. Using an estimate of 887 tons of oil per kilometer of shoreline, Gundlach and Hayes estimated that a total of 63,828 metric tons of oil were deposited along the coastal zone. That was approximately one-third of the total amount of oil lost from the Amoco Cadiz. The remaining two-thirds must be accounted for by evaporative processes, oil masses remaining on the water surface, sinking to the bottom and/or mixing into the water column. One month later, a total of 10,310 metric tons of oil could be accounted for on the beaches. This was an 84% decrease in the oil found along the shore during the first visit, although more coastline was covered due to a wind shift on April 2, 1978, causing the oiling of previously clean coastal areas south of the wreck site. A total of 72 kilometers of coastline were covered immediately following the spill during the first two weeks, and after one month this had increased to 213 kilometers of lightly oiled beaches and 107 kilometers of heavily oiled beaches. The continued apparent loss of oil from the shores was attributed to a combination of natural cleaning processes due to

126  FATE AND WEATHERING OF PETROLEUM SPILLS

heavy surf and tidal activities and a very active cleanup program. Considerable contamination of interstitial water occurred following this spill and significant quantities of oil no doubt remained on the beaches even though they appeared to be clean. This is obviously another sink which can receive much of the oil which is unaccounted for or otherwise believed to be removed.

Slightly different estimates from those of Hayes and Gundlach for the mass balance of oil following the Amoco Cadiz spill were completed by the French press. Figure V-1 shows an estimate of the fate of the oil from the Amoco Cadiz as reported in the local press and later attributed to some of the local French officials. It indicates substantial losses to the sea and most likely, to beach sands. However, considerable amounts of oil remain unaccounted for with either approach.

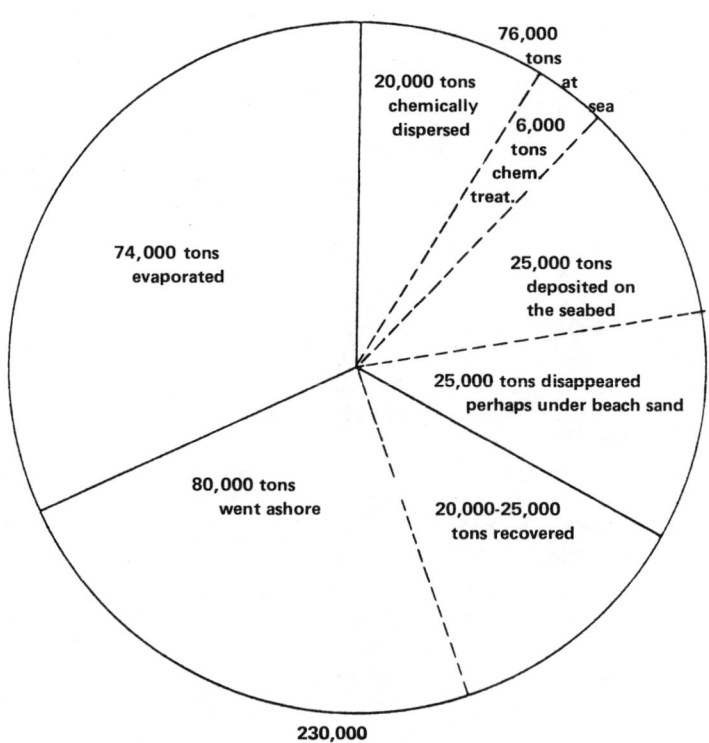

Figure V-1. Estimated fate of oil from Amoco Cadiz (source: local French newspapers).

## OPEN OCEAN MASS BALANCE

Butler et al. (39) have cited work by Morris (165), on pathways for the environmental fate of crude oil spilled on the open ocean. They estimated that evaporation would remove approximately 25% of the materials on a time scale of one to ten days. Dissolution could remove an additional 5% on that same time scale. Photochemical reactions occur over longer periods of time (10-100 days) and can remove approximately 5% of the spilled oil. Microbial degradation on a time scale of 50-500 days was estimated to remove an additional 30% of the material, and disintegration and sinking on a 100 to 1,000 day time scale removed a final 15%. The residual materials, which constituted approximately 20% of the total, were then estimated to remain on the surface as tar balls in excess of hundreds of days.

In an actual spill situation in the Gulf of Mexico approximately 50% of the oil lost due to a well blowout could be accounted for (154, 156, 157). In 1970 a $34^0$ API gravity crude petroleum well blew out from a production platform in the Gulf of Mexico over a three-week period. It was estimated that 25-30% of the oil was lost due to evaporation of the lower molecular weight compounds. An additional 10-20% of the oil was removed from the sea surface by skimming and other recovery techniques; less than 1% was believed to be dissolved in seawater; and less than 1% was deposited in the sediment within an 8 kilometer radius of the well. Detergents were used in that cleanup operations and also in the water sprayed on the burning platform to help disperse the oil. These considerations account for approximately 50% of the oil released. Evidently, the remaining oil became emulsified and dispersed to undetectable levels (less than 1 ppm) or became biodegraded and/or photooxidized.

As part of a larger program assessing the environmental impact of treated versus non-treated oil spills, Cornillon et al. (54) developed a fate model which tracks both the surface and subsurface oil concentrations using computer techniques. This model was then applied to a simulated 34,840 metric ton spill of a Number 2 type oil on George's Bank, and the concentration of oil in the water column and surface slick trajectory were predicted as a function of time for chemically treated and untreated spills occurring during the months of April and December. Figures V-2 and V-3 present mass balance considerations of untreated and treated oil spills as a function of time. Because the theoretically spilled oil contained large amounts of the lighter fractions, the evaporation rate was very rapid with

128  FATE AND WEATHERING OF PETROLEUM SPILLS

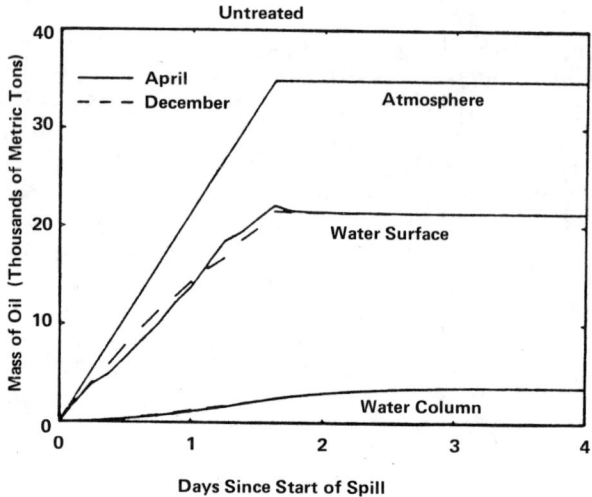

Figure V-2. Mass balance of untreated spilled oil, April and December (from Cornillion et al., 54). Reprinted with permission of the authors and the Oil Spill Conference Office, Washington, DC.

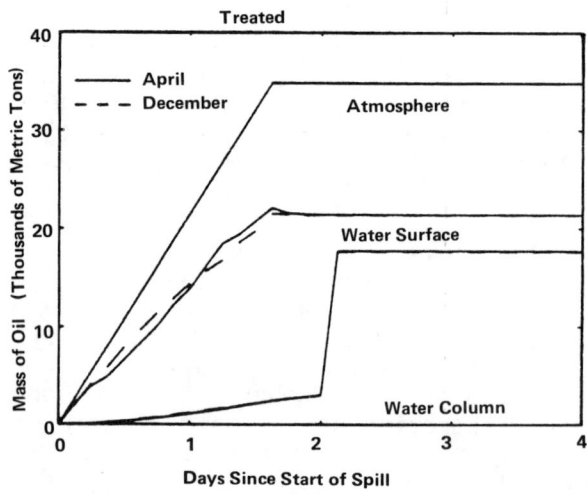

Figure V-3. Mass balance of treated spilled oil, April and December (from Cornillion et al., 54). Reprinted with permission of the authors and the Oil Spill Conference Office, Washington, DC.

approximately 41% of the oil being evaporated within several hours after its release. For the untreated spill approximately 10% of the oil was slowly dispersed into the water column and Figure V-2 illustrates that approximately 49% of the oil mass was predicted to remain on the surface slick. The only differences between the April and December spills in the untreated case were slight variations in the evaporation rates, caused by different simulated wind patterns and a small seasonal temperature difference. The ultimate amount of oil lost to evaporation during both months remains unchanged.

The main difference between the predicted behavior of the treated and untreated situation was the rapid increase of oil in the water column as shown after the addition of emulsants two days after the spill. The final mass balance in Figure V-3 for the treated oil shows 39% evaporated, 57% dispersed in the water column, and 10% of the oil remaining on the surface. The computer model predicted that 2% less oil would be evaporated in the treated case because of the fact that the last amount of spilled oil released was theoretically treated at the time of release and therefore was dispersed before all of its lighter fractions could be evaporated. Again, seasonal effects are only noted in the evaporation rates.

The computer model also predicted the relative transport between the surface slick and oil dispersed in the water column and Figures V-4 and V-5 show trajectories and areas of spill concentrations during the months of April and December, respectively. During the month of April, the simulated wind patterns clearly show that the surface slick moves in a different direction than the subsurface dissolved and dispersed oil. The solid elipses show the 50 $\mu g/l$ contour lines for the untreated case 5, 10, 20 and 30 days following the start of the spill, and the dashed ellipse shows similar plots for the treated oil.

This paper is one of the few which attempt to provide mass balance calculations on oil released in an open ocean situation. To date, very little actual data exist on such spills. From the model, it was clear that in order to improve predictive capabilities a number of areas must be further examined. Specifically, a better understanding of oil spill entrapment, spreading and emulsification, both in the presence of chemical treatment and for untreated spills, is needed. Also, their model did not adequately address small scale circulation dynamics which would provide a more realistic approximation of the critical advective and diffusive transport processes. The model also was not

130   FATE AND WEATHERING OF PETROLEUM SPILLS

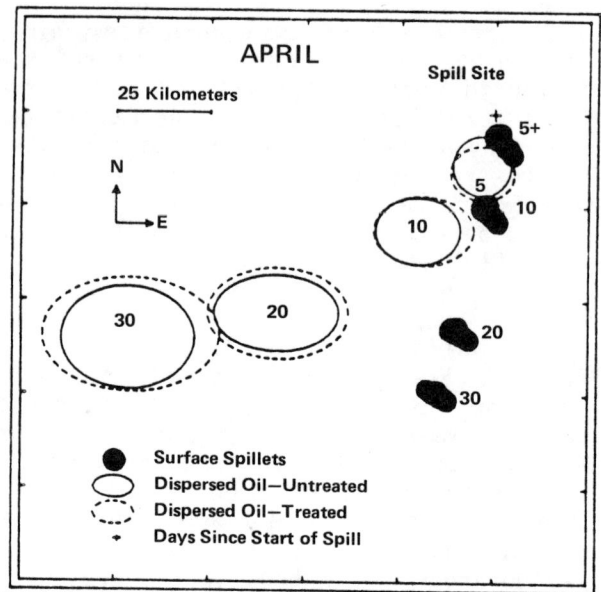

Figure V-4. Trajectories and areas of the surface spillets and the dispersed oil (50-ppb contour) for the treated and untreated spill in April (from Cornillion et al., 54). Reprinted with permission of the authors and the Oil Spill Conference Office, Washington, DC.

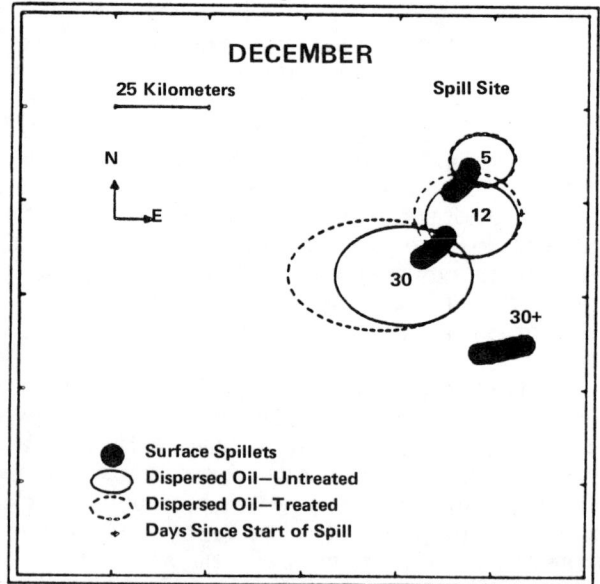

Figure V-5. Trajectories and areas of the surface spillets and the dispersed oil (50-ppm contour) for the treated and untreated spill in December (from Cornillion et al., 54). Reprinted with permission of the authors and the Oil Spill Conference Office, Washington, DC.

particularly applicable to near-shore processes, which would include stranding of the oil on the shoreline and/or deposition of oil in coastal marshes. Nevertheless, with appropriate use in open ocean spills, this model could help to better the understanding of the complex dynamics of oil spills and help to refocus research to previously unrecognized problem areas.

MASS BALANCE IN AN ESTUARINE ENVIRONMENT

In an attempt to estimate rates of transport and the fate of petroleum in an estuarine environment, Gearing et al. (85) studied the chemical fate and alterations of a water accommodated fraction of Number 2 fuel oil in control ecosystem tanks at the Marine Ecosystems Research Laboratory at the University of Rhode Island. Their hydrocarbon budget results indicated that the main loss of oil was to the atmosphere, although the oil used in their work was a Number 2 fuel oil, so most of the aliphatic and aromatic compounds had volatilities greater than $n-C_{22}$.

The hydrocarbon concentrations were measured in the water over time and from these data an overall rate was calculated for the oil loss from the sum of evaporation, biodegradation, and adsorption of oil onto particulates. To determine the rates of more individual processes, the authors stated that it would be necessary to look at more parts of the system and to look at rates of loss of individual compounds.

In their ecosystem enclosures, particulates were observed to adsorb approximately 15% of the water accommodated fraction of the fuel oil and these particulate oil globules then resulted in the sedimentation of approximately 7-16% of the added oil. The oil found in the sediments was depleted in the lower molecular weight aromatics (up to 3-ring compounds) reflecting the differential evaporative losses and solution of these materials. Slightly different concentrations of hydrocarbons were found in the water column, reflecting the differential solubilities of some of the components. For example, in an oil/water dispersion generated by gentle swirling, the aromatic components constituted 30-40% of the total water dispersed/dissolved material whereas they were approximately only 20-25% of the original oil. If rapid or vigorous shaking was used the dispersion, small droplets with micelles in the 0.3-6 micron diameter range occurred, and these droplets had a composition closer to the original oil. In the gently swirled situation no preferential fractionation of the less soluble aliphatics into the water was observed. In the sediments, which acted as the ultimate sink for some fraction of the oil, 2-34% of the aliphatic, alicyclic and larger aromatic hydrocarbons were found, in contrast to only 0.1% of the naphthalenes and methylnaphthalenes which were the principal aromatic components in the starting Number 2 fuel oil. These results indicate that there are substantial differences in the mechanisms and/or rates by which aliphatic and aromatic compounds move through the marine ecosystems, and as in other mass balance studies, a significant portion of the oil is lost to evaporation and/or otherwise unaccounted for. It was also noted in these studies that the Number 2 fuel oil hydrocarbons were most concentrated in a surface flocculant layer which was mixed to a depth of 5 cm by physical and biological processes. Due to difficulties in sampling this layer, it has been overlooked, or not sampled adequately in the past during oil spill situations, and it should perhaps be considered in greater detail when trying to establish overall hydrocarbon budgets.

MASS BALANCE OF OIL RELEASED IN ICE COVERED AREAS

Perhaps the best approach at mass balance calculations for oil spilled on or under ice in an actual spill situation was reported by Baxter et al. (19) following the Bouchard #65 oil spill in Buzzard's Bay in January, 1977. On January 28, 1977, the barge Bouchard #65 ran aground in Buzzard's Bay, Massachusetts, spilling 81,146 gallons of Number 2 heating oil into the ice-covered bay. The oil was swiftly carried by currents under the ice and through crack systems where it collected in rafted ice, rubble fields, and pressure ridges. The oil was contained in this ice matrix until the ice began to break up eight days after the spill. At this time the oil was released from the ice in the form of thin sheens and oily flows, and transported 27 miles by currents through the Cape Cod Canal into Cape Cod Bay. Cleanup operations recovered approximately 18.3% of the spill; 84% of this recovered oil by direct suction from pools near rafted ice. A burn was attempted on one 4,000 gallon pool in a ridge system and it is estimated that 50% of that oil was burned off.

In an attempt to complete an oil budget for the spill, a series of aerial, photo mosaics designed to document the locations of all the major concentrations of oil visible on the surface in the ice and water was completed. The photographic record was combined with sampling of oiled ice surfaces and slush to determine the percent oil/ice composition by volume and depth of penetration. Through the photographic observations and field notes, general classifications of visibly yellowed ice were separated. These included the following:

1. deep oil pools: These were pools of oil ranging in depth from 0.1 to 0.15 meters and they were generally associated with rafted ice. These pools were the primary sites of oil cleanup operations.

2. shallow oil pools: Brash ice shallow pools ranging from from 0.01 to 0.04 meters.

3. heavy oil concentrations in active ice: This was darkly stained ice amid the brash ice and small flows in areas that were heavily oiled in contained pools. Typically, oil penetration in this ice measured 0.05 meters with an average concentration of 5% by volume.

4. medium oil concentrations in active ice: In this case the ice surface was clearly saturated with oil but it was lighter in color than the heavy concentrations considered in 3 above. Oil penetration reached 0.05 meters in concentrations of 1% by volume.

5. light oil concentrations in active ice: In this case the oil was barely visible from air and the ice was not uniformly saturated. Oil penetration reached a depth of 0.05 meters in concentrations of approximately 0.5% by volume.

6. oil on the surface of ice: This included all oiled flows whether oiled by barge dumping, cleanup operations or wind-blown oil. Depth of penetration ranged from 0.001 to 0.003 meters in concentrations of 50% by volume. This was primarily pooled oil.

From the photomosaics each square meter of colored ice was divided into one of the six categories described above and best estimates were made on the pool depth oil penetration and percent oil concentration at that depth. The results of this effort are presented in Table V-1.

Most of the data are self-explanatory, however, the notation "ice surface involved" with a "2" indicates that more than one surface of the ice was covered. The amount of oil lost by evaporation from each type of oil is presented in Table V-2. As discussed in another section, approximate loss due to weathering ranged from 3.7 to 47%, depending on how the oil was distributed (i.e., on the surface of ice or in deep pools or under ice hummocks). That is, weathering losses ranged from 12-13% for oil in deep and shallow pools and went as high as 47% for lightly oiled areas on ice exposed to the air.

The data in Table V-2 illustrate that approximately 12,000 gallons or 15% of the oil was lost to the atmosphere. Of the six categories denoting the oil distribution after release, the largest amount of oil, 19%, was contained in the deep pools. The combination of the deep and shallow pools contained approximately 45% of the total spilled oil. Due to this distribution, it was determined that pumping was the best mechanism for cleaning up the oil following the accident.

By direct suction of pooled oil, approximately 12,886 gallons were recovered. The burning of oil trapped within the ice is estimated to have removed a total of 1,500-2,000 gallons; the remainder of the oil was eventually carried out of Buzzard's Bay on the ice flows and released in Cape Cod Bay. These oily ice flows could not be contained by conventional mechanisms and from this standpoint, it appears that this would be a mechanism for long-range transport of oil which could be important in the Arctic.

In conclusion, these attempts at determining mass balance budgets for spilled oil are only a few examples of calculated distributions of oil in the environment. They represent some of the better studies which have been completed at this time. With the exception of the Buzzard's Bay spill of _Bouchard_ #65, however, the majority of the spilled oil in most situations cannot be traced due to evaporative losses or sinking.

Table V-1. Oil budget for Bouchard #65 oil spill, February 2-4, 1977 (from Baxter et al., 19). Reprinted with permission of the NOAA/Science Applications, Inc. Program, Boulder, CO.

| Type of Oil/Ice Configuration | Area from Mashnee Island to Wings Neck Tower (m²) | Area from Wings Neck tower to Cleveland Ledge (m²) | Total Area (m²) | % Saturation | Ice Surface Involved | Depth of Saturation (m) | Volume (liters) | Volume (gallons) | % of Total |
|---|---|---|---|---|---|---|---|---|---|
| Deep oil pools in rafted ice | 400 | 300 | 700 | 100 | - | 0.13 | 91,000 | 24,000 | 29 |
| Shallow oil pools in brash ice and small floes | 1,500 | 300 | 1,800 | 100 | - | 0.025 | 45,000 | 12,000 | 14 |
| Heavy oil concentrations | 7,400 | 2,400 | 9,800 | 5 | 2 | 0.05 | 49,000 | 13,000 | 16 |
| Medium oil concentrations | 26,600 | 3,200 | 29,800 | 1 | 2 | 0.05 | 30,000 | 8,000 | 10 |
| Light oil concentrations | 28,700 | 3,800 | 32,500 | 0.5 | 2 | 0.05 | 16,000 | 4,000 | 5 |
| Oil on ice surface | 2,100 | 12,600 | 14,700 | 50 | - | 0.003 | 22,000 | 6,000 | 7 |
| Burn site (heavy) oil concentration | - | - | 5,600 | 5 | - | 0.05 | 14,000 | 4,000 | 5 |
| Evaporation losses (see Table 3.2) | - | - | - | - | - | - | 44,000 | 12,000 | 14 |
| TOTALS | ~66,700 | ~22,600 | 94,900 | | | | 311,000 | 83,000 | 100 |

Table V-2. Weathered losses of oil in January 1977 (from Baxter et al., 19). Reprinted with permission of the NOAA/Science Applications, Inc. Program, Boulder, CO.

| Type of Oil/Ice Involvement | Approximate Total Percent Loss | Total in Each Type of Oil/Ice (gal) | Weathered Losses (gal) |
|---|---|---|---|
| Deep Oil Pools | 13 | 24,000 | 3,100 |
| Shallow Oil Pools | 12 | 12,000 | 1,400 |
| Heavy Oil Concentrations | 15 | 13,000 | 2,000 |
| Medium Oil Concentrations | 19 | 8,000 | 15,00 |
| Light Oil Concentrations | 47 | 4,000 | 1,900 |
| Oil on Ice Surface | 31 | 6,000 | 1,900 |
| Burn Site (Heavy Concentration) | 15 | 4,000 | 600 |
| Totals | | 70,000 | 12,400 |
| | | | or ∼12,000 |

## VI. PROSPECTUS

To accurately assess the environmental effects resulting from a petroleum spill, a thorough examination of the chemical and physical changes occurring with weathering of the petroleum must be undertaken. Crude petroleum is a complex mixture of homologous series of hydrocarbon compounds; nitrogen-, sulfur-, and oxygen-containing species; as well as certain trace metals, present in the elemental form, as salts, or as organometallic complexes. A systematic approach is needed for identification and quantification of these components (for crudes and petroleum products) which parallels an identification and classification scheme for the chemical and physical changes occurring with weathering (biotic and abiotic). There is also a need for better definition of biogenic hydrocarbon input and the baseline levels of hydrocarbons in various marine environments. This is very important when assessing petrogenic inputs into a particular ecosystem, as well as when determining the fate of petroleum components following a spill.

In this review, a rather strong emphasis has been placed upon microbial degradation of whole petroleum and petroleum components. This is not to infer that microbial metabolism is the most important factor in the overall process of weathering. It was intended here to emphasize the ubiquitous nature of hydrocarbonoclastic microbial species, the vast array of petroleum components which can be metabolized by these microbes, and the reaction mechanisms and pathways that are involved. To fully assess the actual importance of microbial populations in the fate of a petroleum spill, controlled field studies are needed to determine the extent of petroleum degradation by mixed indigenous microbial populations, as well as the response time to inputs of petroleum in the localized environment. The extent of degradation should be considered as functions of various abiotic environmental parameters or conditions, such as: temperature, salinity, pH, etc., as well as any seasonal fluctuations. In such studies, rate experiments should be properly designed with numerous sampling points at the beginning of the experiments to better characterize the order and value of the determined rate constants. In determining the response time of indigenous microbial populations to petroleum inputs, studies should be designed to determine the rates and extent in shifts of heterotrophic population composition towards a greater percentage of hydrocarbonoclastic species.

More intensive laboratory studies utilizing both simulated environments and coastal ecosystem enclosures are needed to assess chemical changes occurring in petroleum hydrocarbon degradation, and alterations of non-hydrocarbon components due to microbial and abiotic factors. Realistic concentrations of oil should be used in such studies. State-of-the-art techniques such as GC/GCMS, high pressure liquid chromatography (HPLC), IR spectroscopy, etc., should be utilized to determine chemical alterations with time as influenced by the various selected abiotic and biotic parameters.

It appears that evaporation is a primary process for the removal of petroleum hydrocarbons from open ocean and near-shore spills. Direct chemical measurements are needed on air samples above slick surfaces to document the type of hydrocarbons lost and the associated rates. To date, very little information is available on ambient air concentrations of volatile hydrocarbons after spill situations.

More thorough examinations of photochemical oxidation of petroleum compounds are needed to assess the relative importance of this process in overall weathering. It appears that a rather complex relation exists between slick thickness and the wavelengths within the solar spectrum that are most active in initiation of the free-radical chain process. This relation should be investigated with respect to the range of solar intensities (quantitative and qualitative aspects) expected to be found in various environments and weather conditions.

Toxicity studies are needed to determine any detrimental effects of both the dissolved and particulate-sorbed petroleum components towards major types of marine organisms, as well as microbial populations. Also, it appears that the use of chemical dispersants can enhance toxicities in certain situations, especially when studied in conjunction with photochemical processes.

Research needs should be intensified in areas that are more susceptible to detrimental effects from spilled oil, such as estuarine and marsh areas, as opposed to higher energy coasts. The one exception to this would be the examination of oil in near-shore bottom sediments and bottom currents in high energy zones, where oil in the bottom sediments may not be rapidly moved, or where oil may be deposited in the near-shore sediments after it has been removed from the intertidal/sandy beach zone.

To improve mass balance calculations better estimates of evaporative losses are required, as noted above, and interstitial waters should be more intensively studied in beach areas where cosmetic oil removal suggests that oil pollution is not a long-term problem. The subsurface flocculant layer which occurs below a slick should be more intensively studied, also with regard to improving mass balance calculations.

It appears that the fate of oil in benthic sediments is highly dependent upon the oxygen content in the interstitial waters and that distinctly different deposition breakdown burial scenarios exist, depending on the degree of biological activity at the sediment water interface. Further work is necessary to determine if residues are being isolated, redistributed or degraded once they are introduced into the sediments.

REFERENCES

1. Aalund, L. R. 1976. Wide variety of world crudes gives refiners range of charge stocks. Oil Gas J. 74(13):87-122; 74(15);72-8; 74(17):112-26; 74(19):85-94; 74(21):80-7; 74(23):139-48; 74(25):137-52; 74(27):98-108.

2. Alpine Geophysical Associates, Inc. 1971. Oil pollution incident, Platform Charlie, Main Pass Block 41, Field, Louisiana, Project. 15080FTU. Water Pollution Control Research Series. Environmental Protection Agency, Washington, D. C., 134 pages.

3. Anderson, J. W., J. M. Neff, B. A. Cox, H. E. Tatem, and G. H. Hightower. 1974. Characteristics of dispersions and water-soluble extracts of crude and refined oils and their toxicity to estuarine crustaceans and fish. Mar. Biol. (Berl.) 27:75-88.

4. Atlas, R. M. 1973. Fate and effects of oil pollutants in extremely cold marine environments. AD 769895. National Technical Information Service, U. S. Dept. of Commerce, Springfield, Va., p. 13. Cited In: P. C. Deslauriers. 1975. Oil pollution in ice infested waters (A survey of recent development) p. 17. Unpublished manuscript. 9104 Red Branch Road, Columbia, Md.

5. _____ and R. Bartha. 1972. Biodegradation of petroleum in seawater at lower temperatures. Canadian Journal of Microbiology 18(12):1851-1855.

6. _____ and _____. 1973. Stimulated biodegradation of oil slicks using oleophillic fertilizers. Environmental Science and Tech. 7(6):538-541.

7. _____ and _____. 1973. Fate and effects of polluting petroleum in the marine environment. Residue Review 49:49-65.

8. _____. 1975. Effects of temperature and crude oil composition on petroleum biodegradation. Applied Microbiology 30(3): 396-403.

9. _____. 1977. Studies on petroleum biodegradation in the Arctic. In: D. A. Wolfe (ed.), Fate and effects of petroleum hydrocarbons in marine organisms and ecosystems. Pergarmon Press, Inc.

10. Attaway, D., J. R. Jadamec, and W. McGowan. 1973. Rust in floating petroleum found in the marine environment. Unpub. manuscript, U. S. Coast Guard. Cited in: Figs. 7 and 8 of J. N. Butler, B. F. Morris, and J. Sass. 1973. Pelagic tar from Bermuda and the Sargasso Sea. Bermuda Biol. Stn. Spec. Publ. 10, p. 23.

11. Ayers, R. C., Jr., H. O. Jahns, and J. L. Glaeser. 1974. Oil spills in the Arctic Ocean: extent of spreading and possibility of large-scale thermal effects. Science 186:843-5.

12. Baier, R. E. 1972. Organic films on natural waters: their retrieval, identification, and modes of elimination. Journal of Geophysical Research 77(27):5062-5075.

13. Baptist, J. N., R. K. Gholson, and M. J. 1963. Hydrocarbon oxidation by a bacterial enzyme system I. Products of octane oxidation. Biochimica et Biophysica Acta, 69:40.

14. Barber, F. G. 1970. Report of the Task Force: Operation Oil (Cleanup of ARROW oil Spill in Chedabucto Bay). Vol. 3, p. 35-54. Ministry of Transport, Ottawa. In: Proceedings of 1975 Conference on Prevention and Control of Oil Pollution, p. 500. American Petroleum Institute, Washington, D. C.

15. _____. 1971. An oiled Arctic shore. Arctic 24:229.

16. _____. 1971. Oil spilled with ice: Some qualitative aspects, p. 133-137. In: Proceedings of 1971 Joint Conference on Prevention and Control of Oil Spills. American Petroleum Institute, Washington, D. C.

17. Bartha, R. and R. M. Atlas. 1973. Biodegradation of oil in seawater: limiting factors and artificial stimulation. In: D. G. Ahearn and S. P. Meyers (eds.), Microbial degradation of oil pollutants, Louisiana State Univ. Center for Wetland Resources, Pub. No. LSU-SG-73-01.

18. Bassin, J. J., T. Ichiye. 1977. Flocculation behavior of suspended sediments and oil emulsions. Journal of Sedimentary Petrology 47(2):671-677.

19. Baxter, B., P. C. Deslauriers, and B. J. Morson. 1977. The BOUCHARD #65 oil spill, January, 1977, MESA Special Report, National Oceanic and Atmospheric Association, Boulder, Colorado.

20. Beckett, Capt. C. J. 1979. The grounding of the IMPERIAL ST. CLAIR - a case history of contending with oil in ice, p. 371-376. In: Proceedings of the 1979 Oil Spill Conference (Prevention, Behavior, Control, Cleanup), 19-22 March 1979, Los Angeles, Calif.

21. Berridge, S. A. et al. 1968. The properties of persistent oils at sea, p. 2-11. In: Peter Happle (ed.), Scientific aspects of pollution of the sea by oil. Proc. of a symposium.

22. _____. 1968. The formation and stability of emulsions of water in crude petroleum and similar stocks. Journal of the Inst. of Petrol. (London) 54(539):333-357.

23. Bieri, R. H., V. C. Stamoudis, and M. K. Cueman. Chemical investigations of two experimental oil spills in an estuarine ecosystem, part II, p. 693-698. In: Proceedings of the 1979 Oil Spill Conference (Prevention, Behavior, Control, Cleanup), 19-22 March 1979, Los Angeles, Calif.

24. Bitton, G. et al. 1979. Resistance of bacterial chemotaxis to blockage in petroleum waters. Marine Pollut. Bull. 10(2):48-49.

25. Blumer, M., and J. Sass. 1972. Indigenous and petroleum-derived hydrocarbons in a polluted sediment. Mar. Pollut. Bull. 3:92-3.

26. _____, and _____. 1972. Oil pollution: persistence and degradation of spilled fuel oil. Science 176:1120-2.

27. Blumer, M., M. Ehrhardt, and J. H. Jones. 1973. The environmental fate of stranded crude oil. Deep Sea Res. 20:239-59.

28. Boehm, P. D., and J. G. Quinn. 1973. Solubilization of hydrocarbons by the dissolved organic matter in sea water. Geochim. Cosmochim. Acta 37:2459-77.

29. \_\_\_\_, and \_\_\_\_. 1975. Correspondence to the Editor. Environ. Sci. Technol. 9:365.

30. Boylan, D. B., and B. W. Tripp. 1971. Determination of hydrocarbons in seawater extracts of crude oil and crude oil fractions. Nature 230:44-7.

31. Bratberg, E. 1977. The Bravo blowout. Institute of Marine Research, Norway, Series B, No. 5.

32. Brunnock, J. V., D. F. Duckworth, and G. G. Stephens. 1968. Analysis of beach pollutants. J. Inst. Petrol. 54:310-25.

33. Bruyn, J. 1954. Koninkl. Ned. Akad. Wetenschap. Proc. Ser. C, 57:51.

34. Buckley, J., and A. N. Blair Humphrey. 1979. Fate of dispersed oil in the environment, part II, a boomed oil spill. In: Arctic Marine Oil Spill Program Technical Seminar-Preprints, Edmonton, Alberta, Canada, 7-9 March 1979, James F. McClaren, LTD, Edmonton, Alberta.

35. Buckmaster, J. 1973. Viscous-gravity spreading of an oil slick. Journal of Fluid Mechanics 59:481-491.

36. Burwood, R., and G. C. Speers. 1974. Photo-oxidation as a factor in the environmental dispersal of crude oil. Estuarine Coastal Mar. Sci. 2:117-35.

37. \_\_\_\_, B. F. Morris, and J. Sass. 1973. Pelagic tar from Bermuda and the Sargasso Sea. Bermuda Biol. Stn. Spec. Publ. 10. 346 pp.

38. Butler, J. N. 1975. Evaporative weathering of petroleum residues: the age of pelagic tar. Mar. Chem. 3:9-21.

39. \_\_\_\_, \_\_\_\_, Morris, B. F. and T. D. Sleeter. 1976. The fate of petroleum in the open ocean. p. 287-297. In: Sources, effects and sinks of hydrocarbons in the aquatic environment. The American Institute of Biological Sciences.

40. Button, D. K., P. J. Kinney, D. M. Schell, and B. R. Robertson. Hydrocarbon biodegradation in Alaska's Cook Inlet.

41. Calder, J. 1979. Weathering effects on chemical composition of the AMOCO CADIZ Oil. Presented at the Annual Meeting of the American Association for the Advancement of Science, Houston, Texas.

42(a). Calder, J. A., J. Lake and J. Laseter. 1978. Chemical composition of selected environmental and petroleum samples from the Amoco Cadiz oil spill, p. 21-84. In: The AMOCO CADIZ Oil Spill, NOAA/EPA Special Report.

43. Campbell, W. J., and S. Martin. 1973. Oil and ice in the Arctic Ocean: possible large-scale interactions. Science 181:56-8.

44. Chen, E. C. et al. 1974. Spreading of crude oil on an ice surface. Canadian Journal of Chemical Engineering 52:71-74.

45. Clark, R. C., J. S. Finley, B. G. Patten, D. F. Stefani, and E. E. DeNike. 1973. Interagency investigations of a persistent oil spill on the Washington Coast: animal populations studies, hydrocarbon uptake by marine organisms, and algal response following the grounding of the troopship GENERAL M. C. MEIGS, p. 793-808. In: Proceedings of 1973 Joint Conference on Prevention and Control of oil Spills, American Petroleum Institute, Washington, D. C.

46. _____, _____, _____, and E. E. DeNike. 1975. Long-term chemical and biological effects of a persistent oil spill following the grounding of GENERAL M. C. MEIGS, p. 479-487. In: Proceedings of 1975 Conference on Prevention of Control of oil Pollution, American Petroleum Institute, Washington, D. C.

47. _____ and D. W. Brown. 1977. Petroleum: properties and analysis in biota and abiotic systems, p. 1-89. In: D. C. Mailins (ed.), Effects of petroleum on Arctic and Subarctic marine environments and organisms. Academic Press, New York.

48. _____, and W. D. MacLeod, Jr. 1977. Inputs, transport mechanisms and observed concentrations of petroleum in the marine environment, p. 91-223. In: D. C. Mailins (ed.), Effects of petroleum on Arctic and Subarctic marine environments and organisms. Academic Press, New York.

49. Cobet, A. B., H. E. Guard, M. A. Chatigny. 1973. Considerations in application of microorganisms to the environment for degradation of petroleum products. In: D. G. Ahearn and S. P. Meyers (eds.), Microbial degradation of oil pollutants. Louisiana State Univ. Center for Wetland Resources, Pub. No. LSU-SG-73-01.

50. Conomos, T. J. 1975. Movement of spilled oil as predicted by estuarine nontidal drift. Limnol. Oceanogr. 20:159-73.

51. Conover, R. J. 1971. Some relations between zooplankton and Bunker C oil in Chedabucto Bay following the wreck of the tanker ARROW. J. Fish. Res. Board Can. 28:1327-30.

52. Cook, W. L., J. K. Massey, and D. G. Ahearn. 1973. Degradation of crude oil by yeasts and its effects on Lesbitis reticulatas. In: D. G. Ahearn and S. P. Meyers (eds.), Microbial degradation of oil pollutants. Louisiana State Univ. Center for Wetland Resources, Pub. No. LSU-SG-73-01.

53. Cooney, J. J., and J. D. Walker. 1973. Hydrocarbon utilization by Cladosporium resinae. In: D. G. Ahearn and S. P. Meyers (eds.), Microbial degradation of oil pollutants. Louisiana State Univ. Center for Wetland Resources, Pub. No. LSU-SG-73-01.

54. Cornillon, P. C., M. L. Spaulding, and K. Hansen. 1979. Oil spill treatment strategy modeling for Georges Bank. In: Proceedings of the 1979 Oil Spill Conference (Prevention, Behavior, Control, Cleanup), 19-22 March 1979.

55. Cretney, W. J., C. S. Wong, D. R. Green, C. A. Bawden. 1978. Long-term fate of a heavy fuel oil in a spill-contaminated B. C. coastal bay. Journal of Fisheries Research Board of Canada 35(5):521-527.

56. Cundell, A. M., and R. W. Traxler. 1973. Microbial degradation of petroleum at low temperature. Mar. Pollut. Bull. 4:125-7.

57. Davis, S. J., and C. F. Gibbs. 1975. The effect of weathering on a crude oil residue exposed at sea. Water Res. 9:275-85.

58. Davis, J. B. and R. L. Raymond. 1961. Applied Microbiology, 9:383.

59. Davis, S. J., C. F. Gibbs, and K. B. Pugh. 1977. Quantitative studies on marine biodegradation of oil, III. Comparison of different crude oil residues and effects of seawater source. Environ. Pollut. (13) No. 3:203-215.

60. Dean, R. A. 1968. The chemistry of crude oils in relation to their spillage on the sea, Vol. 2, p. 1-6. In: J. D. Carthy and D. R. Arthur (eds.), The biological effects of oil pollution on littoral communities. Suppl. to Field Studies, obtainable from E. W. Classey, LTD., Hampton, Middx., England.

61. deLappe, B. W., R. W. Risebrough, J. C. Shropshire, W. R. Sisteck, E. F. Letterman, D. R. deLappe, and J. R. Payne. In press. The partitioning of petroleum in related compounds between the mussle Mytilus californianus and seawater in the Southern California Bight. Draft Final Report II-15.0. Intertidal Study of the Southern California Bight, submitted to Bureau of Land Management, Washington, D.C. by the Bodega Marine Laboratory, University of California, Bodega Bay, and Science Applications, Inc.

62. Dewling, R. T., and C. C. Silva. 1979. Impact of dispersant use during the BRAZILIAN MARINE incident, p. 269-76. In: Proceedings of the 1979 Oil Spill Conference (Prevention, Behavior, Control, Cleanup), 19-22 March 1979, Los Angeles, Calif.

63. Dexter, R. N., and S. P. Pavlou. 1978. Mass solubility and aqueous activity coefficients of stable organic chemicals in the marine environment: polychlorinated biphenyls. Marine Chemistry 6:41-53.

64. _____, and _____. 1978. Distribution of stable organic molecules in the marine environment: physical chemical aspects, chlorinated hydrocarbons. Marine Chem. 7:67-84.

65. Dibble, J. T., and R. Bartha. 1976. Effect of iron on the biodegradation of petroleum in seawater. Applied and Env. Microbiology. 31(4):544-550.

66. Dudley, G. 1968. The problem of oil pollution in major oil port. p. 21-29. In: J. D. Carthy and D. R. Arthur (eds.) Biological effects of oil pollution on littoral communities supplemental to Field Studies, Vol. 2, obtainable from E. W. Classey, LTD., Hampton, Middx., England.

67. Dodd, E. N. 1971. The effects of natural factors on the movement, dispersal, and destruction of oil at sea. Cited in: National Academy of Sciences (1975). Petroleum in the Marine Environment, p. 47. Washington, D. C.

68. Douros, J. D., and J. W. Frankenfeld. 1968. Applied Microbiology 16:320.

69. _____, and _____. 1968. Applied Microbiology 16:532.

70. D'Oxouville, L., M. O. Hayes, E. R. Gundlach, W. J. Sexton, and J. Michel. 1979. Occurrence of oil in offshore bottom sediments at the AMOCO CADIZ oil spill site, In: Proceedings of the 1979 Oil Spill Conference, March 19-22, Los Angeles, CA, p. 187-192.

71. Duce, R. A., J. G. Quinn, and T. L. Wade. 1974. Residence time for non-methane hydrocarbons in the atmosphere. Mar. Pllut. Bull. 5:59-61.

72. Dudley, G. 1968. The problem of oil pollution in a major oil port, p. 21-29. In: J. D. Carthy and D. R. Arthur (eds.). The biological effects of oil pollution on littoral communities, Suppl. to Field Studies, Vol. 2, obtainable from E. W. Classey, LTD., Hampton, Middx., England.

73. Ehrhardt, M., and M. Blumer. 1972. The source identification of marine hydrocarbons by gas chromatography. Environ. Pollut. 3:179-94.

74. Eganhouse, R. P., and J. A. Calder. 1976. The solubility of medium molecular weight aromatic hydrocarbons and the effects of hydrocarbon co-solutes and salinity. Geochim. Cosmochim. Acta 40:555-61.

75. Fallah, M. H., and R. M. Stark. 1976. Literature review: movement of spilled oil at sea. Marine Technology Society Journal 10:3-17.

76. Farrington, J. W., P. A. Meyers. 1975. Hydrocarbons in the marine environment, p. 109-36. In: G. S. Eglington (ed.), Environmental chemistry, Vol. 1, The Chemical Society, London.

77. Fay, J. A. 1969. The spread of oil slicks on a calm sea, p. 53-63. In: D. P. Hoult (ed.), Oil on the sea. Plenum Press, New York.

78. _____. 1971. Physical processes in the spread of oil on a water surface, p. 467. In: Proceedings of 1971 Joint Conference on Prevention and Control of Oil Spills, American Petroleum Institute, Washington, D.C.

79. Forrester, W. D. 1971. Distribution of suspended oil particles following the grounding of the tanker ARROW. Journal of Marine Research 29:151-170.

80. Foster, J. W. 1962. Bacterial oxidation of hydrocarbons. In: Osamu Hayaishu, (ed.), Oxygenases, Academic Press, New York.

81. Franks, F. 1966. Solute-water interactions and the solubility behavior of long-chain paraffin hydrocarbons. Nature 210:87-88.

82. Freegarde, M. et al. 1971. Oil spilt at sea: its identification, determination and ultimate fate. In: Laboratory Practice 20(1):35-40.

83. _____, and C. G. Hatchett. 1970. The ultimate fate of crude oil at sea. Interim Report. Admiralty Materials Laboratory, U.K. In: Petroleum in the Marine Environment (1975), p. 48. National Academy of Sciences, Washington, D.C.

84. Garrett, W. D. 1973. The surface activity of petroleum and its influence on the behavior of oil at sea. In: Background papers: Inputs, Fates, and Effects of Petroleum in the Marine Environment, Workshop, p. 451-61. National Academy of Sciences, Washington, D. C.

85. Gearing, J. N., P. J. Gearing, T. Wade, J. G. Quinn, H. B. McCarty, J. Farrington and R. F. Lee. 1979. The rates of transport and fates of petroleum hydrocarbons in a controlled marine ecosystem and

a note on analytical variability, p. 555-564. In: Proceedings of the 1979 Oil Spill Conference (Prevention, Behavior, Control, Cleanup) 19-22 March 1979, Los Angeles, Calif.

86. Gebelein, C. D. 1971. Sedimentology and ecology of a carbonate facies mosaic. Ph.D. Thesis, Brown University, Providence, R.I.

87. Gesser, H. P., T. A. Wildman and W. B. Dewari. 1977. Photooxidation of n-hexadecane sensitized by xanthone. Environmental Science and Tech. 11(6):605-608.

88. Gibson, D. T. and W. K. Yeh. 1973. Microbial degradation of aromatic hydrocarbons. In: D. G. Ahearn and S. P. Meyers (eds.), Microbial degradation of oil pollutants. Louisiana State Univ. Center for Wetland Resources, Pub. No. LSU-SG-73-01.

89. _____, V. Mahadevan, and J. F. Davey. 1974. Bacterial metabolism of para- and meta-xylene: oxidation of the aromatic ring. Journal of Bacteriology, 119(3):930-936.

90. _____, D. M. Jerina, H. Yagi, and H. J. C. Yeh. 1975. Oxidation of the carcinogens benzo(a)-pyrene and benzo(a)anthracene to dihydrodiols by a bacterium. Science 189:295-297.

91. _____. 1976. Microbial degradation of carcinogenic hydrocarbons and related compounds, p. 225-238. In: Sources, Effects and Sinks of Hydrocarbons in the Aquatic Environment, Symposium, American university, Washington, D. C. 9-11 August, 1976.

92. _____. 1977. Biodegradation of aromatic petroleum hydrocarbons. In: Fate and effects on petroleum hydrocarbons in marine ecosystems and organisms NOAA Symposium, 10 Nov., 1976.

93. Giger, W. and M. Blumer. 1974. Polycyclic aromatic hydrocarbons in the environment: isolation and characterization by chromatography, visible, ultraviolet, and mass spectrometry. Anal. Chem. 46:1663-71.

94. Glaeser, J. L. 1971. A discussion of the future oil spill problem in the Arctic, p. 479-484. In: Proceedings of 1971 Joint Conference on Prevention and Control of Oil Spills, American Petroleum Institute, Washington, D. C.

95. Glaeser, J. L. 1971. A discussion of the future oil spill problem in the Arctic, p. 479-484. In: Proceedings of 1971 Joint Conference on Prevention and Control of Oil Spills, American Petroleum Institute, Washington, D. C.

96. _____, and G. P. Vance. 1971. A study of the behavior of oil spills in the Arctic. AD717142. National Technical Information Service. U.S. Dept. of Commerce, Springfield, VA. 60 pp.

97. Gordon, D. C., Jr., P. D. Keizer and N. J. Prouse. 1973. Laboratory studies on the accommodation of some crude and residual fuel oils in sea water. J. Fish Res. Board Can. 30:1611-8.

98. Greenlee, R. W. 1960. Factors in the stability of petroleum emulsions. Amer. Chem. Soc. Div. Petrol. Chem. Preprints, 5(3):133-140.

99. Griffin, L. F. and J. A. Calder. 1977. Toxic effect of water-soluble fractions of crude, refined and weathered oils on growth of a marine bacterium. Applied and Environmental Microbiology, 33(5):1092-1096.

100. Guard, H. E. and A. B. Cobet. 1973. The fate of a bunker fuel in beach sand, p. 827-834. In: Proceedings of 1973 Joint Conference on Prevention and Control of Oil Spills, 34. American Petroleum Institute, Washington, D. C.

101. Guire, P. E., J. D. Friede and R. K. Gholson. 1973. Production and characterization of emulsifying factors from hydrocarbonoclastic yeast and bacteria. In: D. G. Ahearn and S. P. Meyers (eds.), Microbial degradation of oil pollutants, Louisiana State Univ. Center for Wetland Resources, Pub. No. LSU-SG-73-01.

102. Hansen, H. P. 1975. Photochemical degradation of petroleum hydrocarbon surface films on seawater. Marine Chemistry, 3:183-195.

103. Harrison, W., M. A. Winnik, P. T. Y. Kwong and D. Mackay. 1975. Crude oil spills. Disappearance of aromatic and aliphatic components from small sea-surface slicks. Environ. Sci. Technol. 9:231-4.

104. Hayes, M. O., E. R. Gundlach and L. D'Ozouville. 1979. Role of dynamic coastal processes in the impact and dispersal of the Amaco Cadiz Oil Spill (March, 1978), Brittany, France, p. 193-200. In: Proceedings of the 1979 Oil Spill Conference, March 19-22, Los Angeles, CA.

105. Herbes, S. E. and L. R. Schwall. 1978. Microbial transformation of polycyclic aromatic hydrocarbons in pristane and petroleum-contaminated sediments. Applied and Environmental Microbiology, 35(2):306-316.

106. Hess, K. W. and C. L. Kerr. 1979. A model to forecast the motion of oil on the sea, p. 653-654. In: Proceedings of the 1979 Oil Spill Conference (Prevention, Behavior, Control, Cleanup) 19-22 March 1979, Los Angeles, Calif.

107. Hill, I. D. 1967. Microbial oxidation of polynuclear aromatic hydrocarbons. U. S. Patent 3, 318, 781.

108. Horowitz, A. and R. M. Atlas. 1977. Continuous open flow-through system as a model for oil degradation in the Arctic Ocean. Applied and Env. Microbiology, 33(3):647-653.

109. Howard, J. A. and K. U. Ingold. 1966. Absolute rate constants for hydrocarbon auto-oxidation III, -methylstyrene, -methylstyrene, and Indene. Canadian J. of Chemistry, 44:1113-1118.

110. Huang, C. P. and H. A. Elliot. 1977. The stability of emulsified crude oils as affected by suspended particles. In: D. A. Wolfe (ed.), Fates and effects of petroleum hydrocarbons in marine organisms and ecosystems, Pergamon Press.

111. Hutchinson, T. C., J. A. Hellebust, D. Mackay, D. Tain, and P. Krauss. Relationship of hydrocarbon solubility to toxicity in algae and cellular membrane effects. In: Proceedings of the 1979 Oil Spill Conference (Prevention, Behavior, Control, Cleanup), 19-22 March, 1979, Los Angeles, Calif., pp. 541-548.

112. Institute of Petroleum Oil Pollution Analysis Committee. 1974. Marine pollution by oil, Applied Science Publishers, Backing, Essex, England, p. 14-16.

113. Jamison, V. W. et al. 1969. Applied Microbiology, 17:853.

114. ____. et al. 1971. Developments in industrial microbiology. In: Amer. Inst. of Biological Sciences, Garamond/Pridemark Press, Baltimore, MD.

115. Jannash, H. W., K. Eimhjellan, C. O. Warson, and A. Farmanfarmaian. 1971. Microbial degradation of organic matter in the deep sea. Science, 171:672-675, Feb., 1971.

116. ____, and C. O. Wirsen. 1973. Deep-sea microorganisms: in situ response to nutrient enrichment. Science, 180:641-643.

117. Johnson, B. H. and T. Aczel. 1967. Analysis of complex mixtures of aromatic compounds by high-resolution mass spectrometry at low-ionizing voltages. Anal. Chem. 39:682-685.

118. Kaneko, T., G. Roubal, and R. M. Atlas. Bacterial Populations in the Beaufort Sea. Arctic, 31(20:97-107.

119. Kator, H. 1973. Utilization of crude oil hydrocarbons by mixed cultures of marine bacteria. In: D. G. Ahearn and S. P. Meyers (eds.), Microbial degradation of oil pollutants. Louisiana State Univ. Center for Wetland Resources, Pub. No. LSU-SG-73-01.

120. Karrick, N. L. 1977. Alterations in petroleum resulting from physico-chemical and microbiological factors. In: D. C. Malins (ed.), Effects of petroleum on arctic and subarctic marine environments and organisms, Academic Press, New York.

121. Keevil, B. E. and R. O. Ramseier. 1975. Behavior of oil spilled under floating ice, p. 497-501. In: Proceedings of 1975 Joint Conference on Prevention and Control of Oil Pollution. American Petroleum Institute, Washington, D.C.

122. Kinney, P. J., D. K. Button and D. M. Schell. 1969. Kinetics of dissippation and biodegradation of crude oil in Alaska's Cook Inlet, p. 333-340. In: Proceedings of 1969 Joint Conference on Prevention and Control of Oil Spills, American Petroleum Institute, Washington, D.C.

123. Klein, D. A. et al. 1968. Antonie van Leeuwenhoek, 34:495.

124. Klein, A. E. and N. Pilpel. 1974. The effects of artificial sunlight upon floating oils. Water Research, 8:79-83.

125. Kolpack, R. L. 1969. Santa Barbara oil pollution project progress report; marine geology. Mar. Pollut. Bull. 1(18):5-8. (Old Series).

126. Kreider, R. E. 1971. Identification of oil leaks and spills, p. 119-124. In: Proceedings of 1971 Joint Conference on Prevention and Control of Oil Spills, American Petroleum Institute, Washington, D.C.

127. Lacaze, J. C. and O. Villedon de Naide. 1976. Influence of illumination on phytotoxicity of crude oil. Mar. Pollut. Bull. 7:73-6.

128. Larson, R. A., L. L. Hunt, and D. W. Blankenship. 1977. Formation of toxic products from a No. 2 fuel oil by photooxidations, Environmental Science and Tech., 11(5):492-496.

129. Larson, R. A., D. W. Blankenship, and L. L. Hunt. Toxic hydroperoxides: photochemical formation from petroleum constituents, Stound Water Research Center, R.D. 1, Box 512, Anondale, Pa.

130. Leadbetter, E. R. and J. W. Foster. 1960. Arch. Mikrobiol, 35:92.

131. Lee, R. F. and C. Ryan. 1976. Biodegradation of petroleum hydrocarbons by marine microbes. In: J. M. Sharpley and A. M. Kaplan (eds.), Proc. Third International Biodegradation Symp., Applied Science Publishers, London.

132. _____. 1976. Metabolism of petroleum hydrocarbons in marine sediments, p. 334-344. In: Sources effects and sinks of hydrocarbons in the aquatic environment, Symposium, American Univ., Washington, D.C., 9-11 Aug. 1976.

133. _____, and M. Takahashi. 1978. Fate of polycyclic aromatic hydrocarbons in controlled ecosystem enclosures. Env. Science and Tech. 12(7):832-838.

134a. Levy, E. M. 1972. Evidence for the recovery of the waters off the east coast of Nova Scotia from the effects of a major oil spill. Water, Air, Soil Pollut. 1:144-148.

135. Lin, D. L. and B. J. Dutka. 1973. Biological oxidation of hydrocarbons in aqueous phase. Journal WPCF, 45(2):232-239.

136. _____. 1973. Microbial degradation of crude oil and the various hydrocarbon derivatives. In: D. G. Ahearn and S. P. Meyers (eds.), Microbial degradation of oil pollutants, Louisiana State Univ. Center for Wetland Resources, Pub. No. LSU-SG-73-01.

137. Liu, H. T. and J. T. Lin. 1978. Effects of an oil slick on wind waves, pp. 665-674. In: Proceedings of the 1979 Oil Spill Conference (Prevention, Behavior, Control, Cleanup), 19-22 March, 1979, Los Angeles, Calif.

138. Ludwig, H. F. and R. Carter. 1961. Analytical characteristics and oil-tar materials on southern California beaches. J. Water Pollut. Control Fed. 33:1123-39.

139. Mackay, D., P. J. Leinonen, J. C. K. Overall, and B. R. Wood. 1975. The behavior of crude oil spilled on snow. Arctic 28:9-20.

140. _____ and W. Y. Shiu. 1976. Aqueous solubilities of weathered northern crude oils, Bulletin of Environmental Contamination of Toxicology, 15(1):101-109.

141. _____ and _____. The aqueous solubility and air-water exchange characteristics of hydrocarbons under environmental conditions, p. 93-110. In: Chemistry and physics of aqueous gas solutions.

142. Majewski, J. and J. O'Brian. 1974. A kinetic study of a fuel oil undergoing photochemical weathering. Environmental Letters 7(2);145-161.

143. Malinky, G. and D. G. Shaw. 1979. Modeling the association of petroleum hydrocarbons and sub-arctic sediments, p. 621-24. <u>In</u>: Proceedings of the 1979 Oil Spill Conference (Prevention, Behavior, Control, Cleanup), 19-22 March, 1979, Los Angeles, Calif.

144. Marr, E. K. and R. W. Stone. 1961. J. Bacteriology 81:425.

145. Marsh, Lt. G. D., L. A. Schultz, and F. W. DeBord. 1979. Cold regions spill response, p. 355-358. <u>In</u>: Proceedings of the 1979 Oil Spill Conference (Prevention, Behavior, Control, Cleanup), 19-22 March, 1979, Los Angeles, Calif.

146. Martin, S. and W. J. Campbell. 1974. Oil spills in the Arctic Ocean: extent of spreading and possibility of large-scale thermal effects (response). Science 186:845-46.

147. _____, P. Kauffman and P. E. Welander. 1976. A laboratory study of the dispersion of crude oil within sea ice grown in a wave field. Unpublished Report. University of Washington, Dep. Oceanogr. Spec. Rep. 69. 34 pp.

148. Maykut, G. A. and N. Untersteiner. 1971. Some results from a time-dependent thermodynamic model of sea ice. Journal of Geophysical Research, 76(6):1550-1575.

149. Mayo, D. W., D. S. Page, J. Cooley, E. Sorenson, F. Bradley, E. S. Gilfillan, F. A. Hanson. 1978. Weathering characteristics of petroleum hydrocarbons deposited in fine clay marine sediments, Searsport, Maine, Journal of Fisheries Research Board of Canada, 35(5):552-565.

150. McAucliffe, C. 1963. Solubility in water of $C_1$-$C_9$ Hydrocarbons. Nature 200:1092-93.

151. _____. 1966. Solubility in water of paraffin, cycloparaffin, olefin, acetylene, cycloolefin, and aromatic hydrocarbons. J. Phys. Chem. 70:1267-75.

152. _____. 1969. Solubility in water of normal $C_9$ and $C_{10}$ alkane hydrocarbons. Science 163:478-79.

153. _____. 1969. Determination of dissolved hydrocarbons in subsurface brines. Chem. Geol. 4:225-33.

154. _____, A. E. Smalley, R. D. Groover, W. M. Welsh, W. S. Pickle, and G. E. Jones. 1975. Chevron Main Pass block 41 oil spill: chemical and biological investigation, p. 555-556. In: Proceedings of 1975 Conference on Prevention and Control of Oil Pollution. American Petroleum Institute, Washington, D.C.

155. _____. 1976. Personal communication with Clark and MacLeod (47).

156. _____. 1977. Dispersal and alteration of oil discharged on a water surface, p. 19-35. In: Douglas A. Wolfe (ed.), Fate and Effects of Petroleum Hydrocarbons in Marine Ecosystems and Organisms, Pergamon Press, Oxford.

157. _____. 1977. Evaporation and solution of $C_2$ to $C_{10}$ hydrocarbons from crude oils on the sea surface, p. 363-372. In: Douglas A. Wolfe (ed.), Fate and Effects of Petroleum Hydrocarbons in Marine Organisms and Ecosystems, Pergamon Press.

158. Mcminn, T. J. and P. Golden. 1973. Behavioral characteristics and cleanup technique of North Slope crude oil in arctic winter environment, p. 263-276. In: Proceedings of 1973 Joint Conference on Prevention and Control of Oil Spills, American Petroleum Institute, Washington, D.C.

159. _____. 1973. Oil spill behavior in a winter arctic environment. Offshore Technology Conf. Preprint no. OTC-1747.

160. Mechalas, B. J., T. J. Meyers and R. L. Kolpack. 1973. Microbial decomposition patterns using crude oil. In: D. G. Ahearn and S. P. Meyers (eds.), Microbial degradation of oil pollutants. Louisiana State Univ. Center for Wetland Resources, Pub. No. LSU-SG-73-01.

161. Meyers, P. A. 1976. Sediments--sources of sinks for petroleum hydrocarbons?. In: Sources, Effects and Sinks of Hydrocarbons in the Aquatic Environment; Proceedings of the Symposium, American university, Washington, D.C., 9-11 August, 1976 (American Institute of Biological Sciences 1977).

162. Meyers, P. A. and J. G. Quinn. 1973. Association of hydrocarbons and mineral particles in saline solutions. Nature 244:23-4.

163. Milne, D. 1950. Character of waste oil emulsions. Sewage Industry Wastes 22:326-330.

164. Monaghan, P. H. and C. B. Koons. 1973. Petroleum in the marine environment: Gulf of Mexico. Gulf Coast Association of Geological Societies Proceedings, Houston, Texas, p. 49. In: Petroleum in the Marine Environment (1975), National Academy of Sciences, Washington, D.C.

165. Morris, B. F. 1976. The environmental fates of petroleum in marine waters (environmentally induced changes on the petroleum). In: Second IOC/WMO workshop on marine pollution (petroleum) monitoring, Monte Carlo, Monaco, 14-18 June, 1976. Unesco document IOC-WMO/MPMSW-II/_2.

166. Morris, R. J. and F. Culkin. 1974. Lipid chemistry of eastern Mediterranean surface layers. Nature 250:640-2.

167. Mulkins-Phillips, G. J. and J. E. Stewart. 1974. Distribution of hydrocarbon-utilizing bacteria in Northwestern Atlantic waters and coastal sediments, Can. J. Microbiol. 20(7): 955-962.

168. _____. 1974. Effect of four dispersants on biodegradation and growth of bacteria on crude oil. Applied Microbiology 28(4):547-552.

169. _____. 1974. Effect of environmental parameters on bacterial degradation of bunker C, crude oils and hydrocarbons. Applied Microbiology 28(6):915-922.

170. Murray, Stephen P. 1972. Turbulent diffusion of oil in the ocean. Limnology and Oceanography 17:651-660.

171. National Academy of Sciences. 1975. Petroleum in the marine environment, Washington, D.C., 107 pp.

172. National Oceanic and Atmospheric Administration: The ARGO MERCHANT Oil Spill, Washington, D.C., March, 1977.

173. Nelson-Smith, A. 1973. Oil pollution and marine ecology. Plenum Press, New York, 260 pp.

174. Nitkowski, M. F., S. Dudley, and J. T. Graikoski. 1977. Identification and characterization of lipolytic and proteolytic bacteria isolated from marine sediments, Marine Pollution Bull. 8(12):276-279.

175. Nixon, A. C. 1961. Autoxidation and antioxidants, vol. II, W. O. Lundbery (ed.), Interscience Publishers, London.

176. NORCOR engineering Research Ltd. 1975. The interaction of crude oil with Arctic Sea ice. Beaufort Sea Tech. Rept 27, Environment Canada, Victoria, B.C., 206 pp.

177. Olivieri, R., P. Bacchin, A. Robertiello, N. Oddo, L. Degen, and A. Tonolo. 1976. Microbial degradation of oil spills enhanced by a slow-release fertilizer, Applied and Environmental Microbiology, 31(5):629-634.

178. _____, A. Robertiello, and L. Degen. 1978. Enhancement of microbial degradation of oil pollutants using lipophilic fertilizers, Marine Pollution Bull. 9:217-220.

179. Ooyama, Jiro and J. W. Foster. 1965. Antonie van Leeuwenhoek 31:45.

180. Outer Continental Shelf Environmental Assessment Program. 1976. Arctic Project Office, Geophysical Institute, University of Alaska, Fairbanks. Arct. Proj. Bull. 9, p. 32.

181. Overton, E. B., J. R. Patel, and J. L. Laseter. 1979. Chemical characterization of mousse and selected environmental samples from the AMOCO CADIZ oil spill, p. 169-174. In: Proceedings from 1979 Oil Spill Conference, Los Angeles, CA, 19-22 March, 1976.

182. Pancirov, R. J. and R. A. Brown. 1975. Analytical method for polynuclear aromatic hydrocarbons in crude oils, heating oils, and marine tissues, p. 103-13. In: Proceedings of 1975 Conference on Prevention and Control of Oil Pollution. American Petroleum Institute, Washington, D.C.

183. Parker, C. A. 1970. The ultimate fate of crude oil at sea; uptake of oil by zooplankton, p. 242. In: P. Hepple (ed.), Water Pollution by Oil, Institute of Petroleum. London.

184. _____, M. Freegarde, and C. G. Hatchard. 1971. The effect of some chemical and biological factors on the degradation of crude oil at sea. In: P. Hepple (ed.), Water Pollution by Oil. Institute of Petroleum, London, p. 237-44.

185. Parker, P. L. and S. Macko. 1978. An intensive study of the heavy hydrocarbons in the suspended particulate matter of seawater, ch. 11 of South Texas Outer-Continental Shelf BLM Study.

186. Perry, J. J. and C. E. Cerniglia. 1973. Studies on the depradation of petroleum by Filamentous Fungi. In: D. G. Ahearn and S. P. Meyers (ed.), Microbial degradation of oil pollutants. Louisiana State Univ. Center for Wetland Resources, Pub. No. LSU-SG-73-01.

187. Phillips, C. R. and D. M. Groseva. 1975. Separation of multi-component hydrocarbon mixtures spreading on a water surface. Separation Science, 10:111-118.

188. Pipel, N. Photo-oxidation of oil films sensitized by Naphthalene derivatives, Inst. of Petroleum Report 75-007, London, England.

189. Rashid, M. A. 1974. Degradation of bunker C oil under different coastal environments in Chedabucto Bay, Nova Scotia. Estuarine Coastal Mar. Sci. 2:137-44.

190. Raymond, R. L., V. W. Jamison, and J. O. Hudson. 1967. Applied Microbiology, 15:857.

191. _____, _____, and _____. 1970. Hydrocarbon cooxidation in microbial systems, Lipids 6(7):453-457.

192. Regnier, Z. R. and B. F. Scott. 1975. Evaporation rates of oil components. Environmental Science and Technology 9:469-472.

193. Reisfeld, A., E. Rosenburg, and D. Gutnick. 1972. Microbial degradation of crude oil. Factors affection the dispersion in sea water by mixed and pure culture. Applied Microbiology 24(3):363-368.

194. Risebrough, R. W., B. W. deLappe, and T. T. Schmidt. 1976. Bioaccumulation factors of chlorinated hydrocarbons between mussels and seawater. Marine Pollution Bulletin 7:225-228.

195. Robichaux, T. J. and H. N. Myrick. 1971. Chemical enhancement of the biodegradation of oil pollution. Third Annual Offshore Technology Conf. Houston, Tx. 1:515-524.

196. _____ and _____. 1972. Chemical enhancement of the biodegradation of oil pollutants. Journal of Petroleum Tech. 24:16-20.

197. Robertson, B., S. Arhelger, P. J. Kinney, and D. K. Button. 1973. Hydrocarbon biodegradation in Alaskan waters. In: D. G. Ahearn and S. P. Meyers (eds.), Microbial degradation of oil pollutants. Louisiana State Univ. Center for Wetland Resources, Pub. No. LSU-SG-73-01.

198. Rosini, F. F. et al. 1975. Selected values of physical and thermodrynamic properties of hydrocarbons and related compounds. In: The American Petroleum Institute, Pittsburgh.

199. Rostad, H. 1976. Behavior of oil spills with emphasis on the North Sea. Report, Continental Shelf Institute, Trondheim, Norway.

200. Rouball, W. T., D. H. Bovee, T. K. Collier, and S. T. Stranahan. 1977. Flow through system for chronic exposure of aquatic organisms to seawater-soluble hydrocarbons from crude oil. Construction and applications. In: Proceedings of 1977 Oil Spill Conference. American Petroleum Institute, Washington, D.C.

201. Scott, G. 1965. Atmospheric oxidation and antioxidants, p. 529. Elsevier Publishing Co., Amsterdam.

202. Scott, B. F., E. Nagy, J. P. Sherry, B. J. Dutka, V. Glooschenko, N. B. Snow, and P. J. Wode. Ecological effects of oil dispersant mixtures in fresh water. In: Proceedings of the 1979 Oil Spill Conference (Prevention, Behavior, Control, Cleanup.) 19-22 March 1979. Los Angeles, Calif.

203. Seifert, W. K. and J. M. Moldowan. 1978. The effect of biodegradation on steranes and terpanes in crude oils. Geochimica et Cosmochimica Acta 43:111-126.

204. Sieburth, J. M. N. 1971. Distribution and activity of oceanic bacteria. Deep-Sea Research 18:1111-1121.

205. Sisler, F. D. and C. E. ZoBell. 1947. Microbial utilization of carcinogenic hydrocarbons. Science 106:521-522.

206. Sivadier, H. O. and P. G. Mikolaj. 1973. Meausrement of evaporation rates from oil slicks on the open sea. In: Proceedings of 1973 Joint Conference on Prevention and Control of Oil Spills p. 475-84. American Petroleum Institute, Washington, D.C.

207. Smith, C. L. and W. G. MacIntyre. 1971. Initial aging of fuel oil films of seawater. In: Proceedings of 1971 Joint Conference on Prevention and Control of Oil Spills p. 457-61. American Petroleum Institute, Washington, D.C.

208. Smith, J. E. 1968. TORREY CANYON pollution and marine life, p. 196. Cambridge University Press, U.K.

209. Soli, G. 1973. Marine hydrocarbonoclastic bacteria. Types and range of oil degradation. In: D. G. Ahearn and S. P. Meyers (eds.), Microbial degradation of oil pollutants. Louisiana State Univ. Center for Wetland Resources, Pub. No. LSU-SG-73-01.

210. Speers, G. C. and E. V. Whithead. 1969. Crude petroleum. In: G. Eglinton and M. T. J. Murphy (eds.), Organic Geochemistry, Methods and Results p. 638-75. Springer-Verlag, New York.

211. Stewart, J. E., R. E. Kallio, D. P. Stevenson, A. C. Jones, and D. Schissler. 1959. Bacterial hydrocarbon oxidation-I. Oxidation of n-hexadecane by a gram-negative coccus. Journal of Bacteriology 78:441-448.

212. _____ and R. E. Kallio. 1959. Bacterial hydrocarbon oxidation-II. Ester formation from alkanes. Journal of Bacteriology 78:726-730.

213. Stolzenback, K. D., O. S. Madsen, E. E. Adams, A. M. Pollack, and C. K. Cooper. 1977. A review and evaluation of basic techniques for predicting the behavior of surface oil slicks. Report no. 222. School of Engineering, MIT.

214. Sutton, C. and J. A. Calder. 1974. Solubility of higher-molecular-weight n-parafins in distilled water and seawater. Environ. Sci. Technol. 8:654-7.

215. _____ and _____. 1975. Reply to correspondence to the editor. Environ. Sci. Technol. 9:365-6.

216. _____ and _____. 1975. Solubility of alkylbenzenes in distilled water and seawater at $25.0^{o}C$. Journal of Chem. and Eng. Data 20(3):320-322.

217. Suess, E. 1968. Calcium carbonate interactions with organic compounds. Ph.D. Thesis, Lehigh University, Bethlehem, Pa.

218. Teal, J. M., C. Burns, and J. Farrington. 1978. Analysis of aromatic hydrocarbons in intertidal sediments, resulting in two spills of Number 2 Fuel Oil in Buzzard's Bay, Massachusetts. Journal of Fisheries Research Board of Canada 35(5):510-520.

219. Thomas, M. L. H. 1977. Long-term biological effects of Bunker-C in the intertidal zone. In: Douglas A. Wolfe (ed.), Fate and Effects of Petroleum Hydrocarbons in Marine Organisms and Ecosystems. p. 238-245. Peramon Press, Oxford.

220. Thompson, J. J., H. J. Coleman, J. E. Dooley, and D. E. Hirsch. 1971. Bumines analysis shows characteristics of Prudhoe Bay crude. Oil Gas J. 69(43):112-20.

221. Topham, D. R. 1975. Hydrodynamics of an oil well blowout. Beaufort Sea Tech. Rept. 33, p. 52. Environment Canada, Victoria.

222. Traxler, R. W. 1973. Bacterial degradation of petroleum materials in low temperature marine environments. In: D. G. Ahearn and S. P. Meyers (eds.), Microbial degradation of oil pollutants. Louisiana State Univ. Center for Wetland Resources, Pub. No. LSU-SG-73-01.

223. Uzuner, M. S. and F. B. Weiskopf. 1975. Transport of oil slicks under a uniform smooth ice cover. Draft report prepared for the Office of Research and Development. U. S. Environmental Protection Agency, Washington, D.C.

224. Van Der Linden, A. C. and G. J. E. Thijsse. 1965. The mechanisms of microbial oxidations of petroleum hydrocarbons. Advances in Enzymology, p. 27, 469.

225. Vandermeulen, J. H., P. D. Keizer and T. Ahern. 1976. Compositional changes in beach sediment-bound ARROW Bunker C: 1970-1976. Unpublished manuscript. Fisheries Improvement Committee. Int. Counc. Explor. Sea CM E:51, 13 pp.

226. Venkatesh, S., H. S. Sakota and A. S. Rizkalla. 1979. Prediction of the motion of oil spills in Canadian Arctic Waters. In: Proceedings of the 1979 Oil Spill Conference (Prevention, Behavior, Control, Cleanup) 19-22 March 1979, Los Angeles, Calif.

227. Walker, J. D. and J. J. Cooney. 1973. Pathway of n-alkane oxidation in Cladosporium resinae. Journal of Bacteriology 115(2):635-639.

228. _____ and R. R. Colwell. 1974. Microbial degradation of model petroleum at low temperatures. Microbial Ecology 1:63-95.

229. _____, _____, Z. Vaituzis, and S. A. Meyer. 1975. Petroleum-degrading chrophyllous Alga Prototheca zopfii. Nature 254:423-24.

230. _____, _____, and L. Petrakis. 1975. Degradation of petroleum by an alga Prototheca zopfii. Applied Microbiology 30(1):79-81.

231. _____, _____, and _____. 1975. Evaluation of petroleum degrading potential of bacteria from water and sediment. Applied Microbiology 30(6):1036-1039.

232. _____, P. A. Seesman, T. L. Herbert, and R. R. Colwell. 1976. Petroleum hydrocarbons. Degradation and Growth Potential of Deep-Sea Sediment Bacteria. Environ. Pollut. (10):89-99.

233. _____ and R. R. Colwell. 1976. Biodegradation rates of components of petroleum. Canadian Journal of Microbiology 22(8):1209-1213.

234. _____ and _____. 1976. Measuring the potential activity of hydrocarbon degrading bacteria. Applied and Env. Microbiology 31(2):189-197.

235. _____ and _____. 1977. Role of autochthonous bacteria in the removal of spilled oil from sediment. Environ. Pollut. (12):51-56.

236. Walsh, F. and R. Mitchell. 1973. Inhibition of bacterial chemoreception by hydrocarbons. In: D. G. Ahearn and S. P. Meyers (eds.), Microbial degradation of oil pollutants. Louisiana State Univ. Center for Wetland Resources, Pub. No. LSU-SG-73-01.

237. Wheeler, R. B. 1978. The fate of petroleum in the marine environment. Exxon Production Research Co. Special Report.

238. Winters, J. K. 1978. Fate of petroleum derived aromatic compounds in seawater held in outdoor tanks and South Texas Outer-continental Shelf Study-BLM, ch. 12. Draft final report.

239. Wolfe, L. S. and D. P. Hoult. 1973. Oil and ice. Tech. Rev. 75(6):45-6.

240. _____ and _____. 1974. Hopeful findings on oil and ice. Tech. Rev. 76(4):73-4.

241. _____ and _____. 1974. Effects of oil under sea ice. Journal of Glaciology 13(69)::473-488.

242. Yamada, K. et al. 1968. Microbial conversion of petro-sulfur compounds. Part I. Isolation and identification of dibenzothiophene-utilizing bacteria. Agri. Biol. Chem. 32:840-5.

243. Youngblood, W. W. and M. Blumer. 1973. Alkanes and alkenes in marine benthic algae. Mar. Biol. (Berl.) 21:163-72.

244. Zajic, J. E. and B. Supplisson. 1972. Emulsification and degradation of Bunker-C fuel oil by microorganisms. Biotechnology and Bioengineering 14:331-343.

245. _____ et al. 1974. Bacterial degradation and emulsification of a no. 6 fuel oil. Environmental Science and Tech. 8(7):664-668.

246. Zephyr, R. R. and B. F. Scott. 1975. Evaporation rates of oil components. Environmental Science and Tech. 9(5):469-472.

247. ZoBell, C. E. 1973. Microbial degradation of oil. Present status, problems and perspectives. In: D. G. Ahearn and S. P. Meyers (eds.), Microbial degradation of oil pollutants. Louisiana State Univ. Center for Wetland Resources, Pub. No. LSU-SG-73-01.

248. Zurcher, F. and M. Thuer. 1978. Rapid weathering processes of fuel oil in natural waters. Analysis and interpretations. Environmental Science and Tech. 12(7):838-843.

# INDEX

abiotic factors/processes 3-54
Abu Dhabi Murbon 115
accomodation 36
acenaphthene-dibenzofurans 61
achlorophyllus alga 55
Achromobacter 55,57
acid 30
　acetic 24
acidity 43
Acinetobacter sp. 55,57
active wavelengths 44
adsorption efficiency 49
advection drift 7
Aeromonas 55,57
aerosols 17,18
agglomeration 44-54,68
albedo values 26
Alcaligenes sp. 57
aldehydes 23,28
　formation 69
algal blooms 68
alkylbenzenes 9,61,82
alkylperoxy radical 24
alkyl-substituted aracyclone 23
n-amylbenzene 61
anthracene 40,49,61,78,89-91
anoxic sediments 106,107
API gradient 9,116,127
approaches to mass balance problems 125-127
aqueous phase 26
Arabian light oil 115
Arabian heavy oil 115
Arctic
　coastal waters 55
　marine environment 63
　oils 124
Argo Merchant 5
aromatic/aliphatic ratio 45
aromatic/n-alkane ratio 105

Arrow 33,34,54,93,108
Arthrobacter 55,57
artificial sunlight 30
Aspergillus versicolor
asphaltenes 63,94,95
　content 43
attractions 49
autocatalytic free-radical 23
autooxidation 24,29

B. healii 57
Bacillus naphthalinicum 57
bacteria 55,59
bacterial slime 43
bacteriocide 66
barnacles 44
　larvae 54
baseline levels 137
beach erosion 96
Beaufort Sea 17,55,56,65
Beijerinekis sp. 57,91
benzene 20,49,61,65,79
　to munonic acid 80
benzocycloparaffins 61
benzodicycloparaffins 61
1,2-benzanthracene (benz(a)- anthracene) 62
benzanthracenes 30,62
benzo(a)anthracene 30,62
benzo(a)pyrene (3,4-benzpyrene) 62
benzoic acid 80
benzopyrenes 30
benzothiophene 12,62
benzperylene 62
binary system 37
biodegradation processes 55,106
biogenic hydrocarbon input 137
biological reworking 104
bioturbation 107

167

168  FATE AND WEATHERING OF PETROLEUM SPILLS

biphenyl  61,91
boiling point  112
bottom waters  44
  current transport  104
Bouchart #65  14,109,111-114
branched to 12 carbons  60
brash ice  111,112
Brasilian Marina  9
BRAS-X-PLUS  9,96
Brevibacterium  55,57
brine channels  110
British Columbia coastal bay  106
1-bromonaphthalene  61
Bunker C  5,12,33,34,44,52,54,95,108,116,122,123
butane  60
butylbenzene  66,80
n-butylbenzene  61
Buzzards Bay  106,109,111,112

$C_3$ to $C_{11}$
$^{14}C$ labeling  52,59,105
Calanus  54
Candida spp.  57,67,78
capillary action  111,114
carbon number  17
carbonyl compounds  30
carboxyl compounds  28,29
carboxylic acids  23,26-28
carcinogen  30
carcinogenic aromatics  63
carcinogenic/toxic polynuclear aromatics  40
catalysts  26
catechol  84
cathechol to acids  80
cell biomass  65
cellular lipid material  69
Cellulomonas galba  57
Cephalosporium acremonium  57
cetyl palmitate  76
chain-terminating reactions  26
Chedabucto Bay, Nova Scotia  33,34,54,93,95
chemical dispersants  29,35,67
chemoreceptors  65
chemotaxis  65
chlorinated hydrocarbons  105
chlorobenzene  61

chloronaphthalenes  61,87
chlorophyll-a  11
chromatogram profiles  29,50
chronic oil pollution  56
chrysenes  30,62
Chukchi Sea  55
cis-dihydrodiol intermediates  78
Cladosporium resinae  55,57,69
clay dispersion  53
coastal ecosystem enclosures  137
coastal zone energy  94
cleanup efficiency  18
cleanup procedures  114,125
colloidal electrolytes  52
colloidal micelles  53
colloids  33,52
comene  61
computerized mass spectrometry  59
condensation reactions  23,28
controlled ecosystem enclosures  59
co-oxidation reactions  69,70'
copepods  54
copolymerizations  23
Corexit
  9527  10
  8666  29,34
Cornybacterium  55,67,74
Corona del Mar, CA  46
co-solutes  37
p-cresol  61
Cunningham elegans  57
current vectors  5
current velocity  112
cyclane  23
cycloalkylbenzenes  61
cyclohexane  9,20,36,60,77
cycloparaffins  77

decalin(decahydronaphthalene)  61
decane  60
deep-sea environment  63
degradation rates  25
dehydrogenation  28
density  22,44,45
  sea ice  109,123
  seawater  123

INDEX 169

density gradient 10
deposition 105
derivatives 57
detergents 9,10,32
detrital mineral 45
dibenzanthracene 62
dibenzothiophene 14,35,62,91,
 102
dibutylphthalate 40
diesel oil #2 6,10
diffusion rates 23,28,42,44,
 130
diffusion theories 5
dihydrodiol formation 77
dihydrodiol intermediates 91
dihydroxynaphthalene 66
dinaphthenebenzenes 61
direct plating 59
dispersion 9,32
dissolution 12,28,35-42
 rates 36,66
dissolved metal ions 26
 organic carbon 11
 organic matter (DOM) 40
 oxygen 11,65
distribution of hydrocarbonoclastic microbes 55-58
diterminal attack 74
dodecane 60
n-dodecane 76
1-dodecanol 9
drift 5,6-9


effects of ice 65
eicosane 40,42,49,60
Ekofish Braus 12,115
electrolytes 40,53
 interaction 52
electrostatic bonding 53
 fields 52
emulsification 7,9,28,23,42,
 43,65,123
 agents 9,32,66
emulsified oil 53
emulsion stability 32
Endomycopsis lypolytica 57
enrighment techniques 59
Enterobacter 55
enzymatic fission 79

enzymes (constitutive/adaptive)
 68
equilibrium vapor pressure 37
ERCO 102
esterification 23,28
esters 24,29
estuarine systems 96,104
ethane 60
ethanol 9
ethyl allophanate 68
ethylbenzenes 61,83
evaporation 7,12-23,138
 preferential loss 133
 rates 16,36,111
excited singlet 26


factors-rates of utilization
 64,65
farnesene 60,63
fatty acid 69
fecal matter 54
ferric hydroxide 64
ferric octoate 68
filamentous fungi
films 53
 spreading 27
 thickness 3,42
Flavobacterium sp. 57,87
fluorene 24
fluorescent spectrometry 34
fractional distillation distribution 116
free oxygen 65
free radicals 23
freshwater ice 114
freshwater microbes 65
fulvic acid 40
Fundi, Bay of 7
fungi 56,59,69


gas chromatography 14,45,49,59,
 97,101,105
 /mass spectrometry 27,29,45,
 59,97,101,102,105,138
gas oil 9
gasoline 16,28
glacially derived sediments 50
glass capillary/gas chromato-

graphy 14,102
flame ionization detection
  47,48,51,100,103
Goleta Point, CA 46,47
grease ice 109,110

halo-toluenes 61
heat of adsorption 42,49
Henry's law 37
heptane 60
heptocyclohexane 60
heptodecane 40,60
heteroatomic compounds 13
heterotrophs 55,56
hexane 60
n-hexadecane 26,27,30,76,83
hexodecane 60
hexadecene-1 60
high littoral zone 43
hopanes 62
HPLC 138
humic substances 49
hummocks 110
Huron, Lake 107
hydrocarbon burdens 105
hydrocarbonoclasts 56,137
hydroperoxide 23
  radicals 24,26
hydrophobic character 33
hydroxy compounds 23

ice meltdown 109,110
ice/water interface 109
illite 49
Imperial St. Clair 114
incident radiation 108
indene 23
inertia 3
infrared spectra 27,50
infrared spectroscopy 49,138
ingestion 54
inoculum 59
interfacial tension 9
interference/enhancement
  degration rates 65-68
intermediate dihydrodiol
  formation 85
intermolecular association 40
interstitial waters 96,97,126,
  138

intersystem crossing 26,27
intertidal sediments 106
ionic catalysts 24
ionic species 40
ionic strength 33,40
Iranian crude oil 32
  light 115
  heavy 115,124
isoalkanes 24
isoprenoid hydrocarbons 17,101,
  105
  /n-alkane ratios 105
isopropyl-substitute aromatics
  23

jet fuel 9
JP-4 107

kaolinite 32,49-51
kerosene 9,16,65
ketones 23,28
kinematic viscosity 20,22,43,
  123
kovat indices 49
Kuwait crude oil 29,43,44,65,
  115,119,123

La Rosa spill 21
l'Aber Wrac'h 7,99,101,102
lead-matrix pumping 110
leads in ice 110
lecithin (chlorine phosphoglyceride) 68
lenses 6,30,31
Lesbistes reticulatas 66
Libya crude oil 43
lipids 35
lithothamnium 102
logarithmic phase bacterial
  growth 63
low-energy environments 104
lube oils 28

macrofauna 107
Marine Ecosystem Research Laboratory (MERL) 12,35,45,105
marine particulates 40

marine phytoplankton communities 29
mass balance calculations 138
  estuarine environment 131, 132
  ice-covered areas 132-136
  open ocean 127-131
McDevit-Long theory 40
mechanical recovery 96
mechanical surface skimming 30
meiofauna 107
membrane transport functions 66
metabolic products 66
  microbial 68-91
meta-cleavage 79
methane 36,55,70
methyl cholanthrene 62
methyl group oxidation 86
methyl ketones 76
2-methylhexane 60,77
methylnaphthalene 14,45,49,61, 87,106
  /n-$C_{18}$ to n-$C_{19}$ ratio 105
methylnaphthoic acid 86
methylphenanthrenes 62
methylstyrenes 23
$MgNH_4PO_4$ 68
micelles 35,36,40
microalgae 29
microbial degradation 55-91, 107,137
microbial inoculum 91
Micrococcus cerificans 57,76
microorganisms in ice 56
minerals 44
models 6,7,33,37,44,51,125
molar volumes 36,38,40
molecular diffusion 20
molecular structure 35
monohydroxylation 80
monomethyl compounds 45
monoterminal oxidation 69
Montmorillonite 49
mousse 12-14,35,42,44,97,101
  balls 97
multiyear ice 109,110
Mycobacterium rhodochrous 57,79
Mytilus californianus 46

naphthalene 9,45,46,61,78,85, 87,123
naphthalenebenzene 61
naphthenephenanthrenes 62
1-naphthol 30,31
1-(2-naphthyl) undecane 61
Narragansett Bay 49
natural oil seep 46
n-$C_{17}$/n-$C_{18}$ ratio 107
n-$C_{18}$/phytane ratio 14
n-$C_{17}$/pristane ratio 14,105
neap berm 96
neopentane 36
nicotinamide adenine dinucleotide (NAD) 69
Nigerian crude oil 32,43
nitrate phosphate salts 67
nitrogen 64,68
Nocardia sp. 57,80,83,84
nonodecane 60
NSO compounds 5,9,32,34,35, 95,105-107,111,112,116, 122,123
nucleic acids 68
nutrients 44
  availability 91
  concentrations 64
  pollution 91
  release 68
Number 2 fuel oil 12,14,15,18, 20,45,49
Number 4 fuel oil 20
Number 6 fuel oil 20

octadecane 60
n-octadecane 76,83
octane 60
octylphosphate 67,68
oil adsorption 49
  Arctic environments 108-114
  budget 125
  droplets 32,54
  estuarine environments 104-108
  film thickness 25
  pancakes 5
  residue 63,93
  spill vulnerability index 97,98
  viscosity 30

oil-in-water dispersion  34,42
  emulsion  32
oil-water interfacial tension
  96
$^{18}O$-incorporation studies  76
olefinic bonds  36
α-olefins  78
oleophilic fertilizers  67
oleophilic nutrients  68
optical density  25
organic acids  24
organic debris  43
organosulfur compounds  26
Orion buoys  7,8
ortho-cleavage  79
ortho-hydroxy derivatives  78
α-oxidations  84
β-oxidations  69,80
oxygen  23
  content (interstitial water)
    139
  requirements  65
  solubility  104
  transport  104
oxygenated materials  44
  polar moieties  53
  radicals  27


PAH
  See polynuclear aromatic
    hydrocarbons
pancake ice  110,112
paraffin oil  9,43
partial vapor pressure  16
particle length  33
particle size distribution  96
particulate fraction  45,46
  hydrocarbon burdens  46
  partitioning  45,46
particulate/subparticulate
  ratio  4
pelagic tar lumps  44
Penicillum spp.  57
pentadecane  12,60
pentane  60
peptizing ions  53
permanent ice  110
peroxides  26,49
Peruvian oil  31

perylene  62
petroleum hydrocarbon utilization
  rates  59-64
pH  32,40,42,67
  ranges  64
Phaeodoctylum tricornutum  29
phenanthrenes  14,40,62,89,91,
  102,123
phenols  23,29,30
phenyl hydantoin  68
phenylacetic acid  80
3-phenyleicosane  61
3-phenylpropionic acid  66,80
phosphate transport  66
phosphorus  64,68
photochemical degradation
  rates  26,108
  oxidation  23-32
  polymerization  44
photosensitizing agents  26-30
phototoxicity  29
physical fragmentation  44
phytane  60,63,97,99,106
  /n-$C_{18}$ ratio  27
plankton  43
polar organic moieties  9,28
polychlorinated biphenyls  37
polymerization  23,30
polynuclear aromatic hydrocarbons  46,52,63,101,105
  alkylnaphthalenes  46
  dibenzothiophenes  46,101
  fluoranthene  46
  fluorene  101
  naphthalenes  101
  phenanthrenes  46,101
  pyrene  46
polysaccharides  67,68
porosity  53,109
porphyrins  43
Port Valdez  52
pour point  43,108,115
preferential utilization  59,63
primary productivity  29
pristane  60,63,97,99,106
  /n-$C_{17}$ ratio  107
propane  60
n-propylbenzene  61
proteins  68
Prototheca zophi  57

Prudhoe Bay crude oil 108,110, 115-117,119
pseudocumene 61
Pseudomonas spp. 55,57,66-68, 74,78-80,84,85,87-90
psychrophilic bacteria 64
purines 68
pyrenes 62,123
pyrimidines 68

quantum efficiencies 24

radical polymerization 28
radical recombination 23
rafted ice 112,113
random motion 7
rate
  of degradation 105
  of evaporation 35
  of slick inoculation 67
  studies 59, 137
recovery techniques 127
refinery fractions 121.
refractory 63
residence times 12,53
residual oil volume 18
residue 9,139
resins 63,94,95
reviews (literature) 115
ridge keels 112
ring fission 80,85,87
ring formation 36

1-salicylic acid 87
salinity 37,42,49,53,65,109
salting-out effects 40
saturation vapor pressure 16
scintillation 59
sea-states 7,17,42
second-order autocatalytical process 23
sediment 56
  bacteria 63
  water interface 106,107
sedimentary regime 97,107,108
sedimentation 44-54
sedimented oil 53
selected ion monitoring GC/MS 46

Serratia marinoruba 57
silicon dioxide 32
sinking 44-54
solubility data 20,30,36,39,41
soluble silica 11
sorption sites 49
south Louisiana crude oil 20, 21,52,53,119
South Texas Outer Continental Shelf Study 45,46
specific gravity 5,43,44,52,94, 95,108,123
spectral densidy of turbulent energy 33
spectrophotometry 23
spill cleanup 30
spill volume 6,18
splash zone 93
spontaneous flocculation 52
spreading 3-6,33,108
  coefficient 3,5,9,30
  detergent influence 9-12
  friction forces 3
  gravitational forces 3
  inertial forces 3
spring tides 96
steranes 62
stop-spreading point 108
subparticulate oil 34
subsurface flocculant layer 139
sulfoxide derivatives 97
  formation 28
  products 26,35,123
sulfur aromatics 91
  content 29,43
  heteroaromatics 123
surface active agents 9,42,43
  charge 32
  products 28
surface wave spectra 6
surfactants 9,28
suspended matter
  carbonates 32
  clays 32
  debris 43
  hydroxides 32
  oil interactions 45
  oxides 32
  particulates 32,34,42,45,54
  sediments 50

tar globules 30,93,127
tar residues 28,30,43
tarball formation 43,44
temperature 42,49,64
  marine/estuarine environment 63
terpanes 62
tertiary C-H bonds 24
tertiary radicals 27
tetradecane 60
n-tetradecane 76
tetralin 61
tetramethylbenzenes 61
thickness gradient 3
thiocyclane-I-oxides 25,29
thiophenes 91
Tiajuana crude oil 31,115
tidal currents 6,96
tidal flats 96,97
tissue matrices 105
toluene 20,36,61,82
toxic components 29,64
  hydrocarbons 20
  hydroperoxides 29
  volatiles 64,66
toxicity studies 138
trace element 116
trace metals 137
trans-dihydrodiol 78
transport mechanism 104
Traube rule 50
tricarboxylic acid (TCA) system 80
tridecane 60
trimethylbenzenes 20,61
trimethylnaphthalenes 14
triplet state 26
turbulent energy spectrum 33
Type I photosensitized oxidation 26

Ubatuba, Brazil 96
ultraviolet light 24,40,108
undecane 60

unresolved complex mixture 29,49,97

van der Waals forces 42,49,52,53
vanadium 26
  content 43
  salts 29
vanadyl porphyrins 29
vapor pressure 16,17,20,36
Venezuelan crude oil 32,34,43
Vibrio sp. 55,57
viscosity 5,9,28,30,34,94,95,108,109,123
viscous oil 42
void volume 111
volatilization 16,108


water density 109
water droplets 43
water temperature 5,42
water-sediment interface 56
wave/ice action 93
wave functions 6
wavelength 25
wax content 43
weathering rates 44,108
wind field vectors 5,7
wind patterns 6,7
wind speed 6

xanthone 26,27
xylenes 61,84


yeasts 55,56,59
  cultures 30


zooplankton 54